T0135755

Diagnostics of Rotor Asymmetries in Inverter-Fed, Variable Speed Induction Machines

Vom Fachbereich Elektrotechnik und Informatik

der Universität Siegen

zur Erlangung des akademischen Grades

Doktor der Ingenieurwissenschaften
(Dr.-Ing.)

genehmigte Dissertation

von

Diplom-Ingenieurin Eva Teresa Serna Calvo

aus Sagunt, Spanien

1. Gutachter: Prof. Dr.-Ing. Mario Pacas
2. Gutachter: Prof. Dr.-Ing. Friedrich W. Fuchs
Vorsitzender: Prof. Dr. rer. nat. Rainer Patsch

Tag der mündlichen Prüfung: 19. Februar 2009

Bibliografische Information der Deutschen Nationalbibliothek

Die Deutsche Nationalbibliothek verzeichnet diese Publikation in der
Deutschen Nationalbibliografie; detaillierte bibliografische Daten sind
im Internet über http://dnb.d-nb.de abrufbar.

ISBN 978-3-8325-2253-7

Logos Verlag Berlin GmbH
Comeniushof, Gubener Str. 47,
10243 Berlin
Tel.: +49 (0)30 42 85 10 90
Fax: +49 (0)30 42 85 10 92
INTERNET: http://www.logos-verlag.de

Acknowledgments

This thesis was written during my stay as research assistant at the Institute of Power Electronics and Electrical Drives of the University of Siegen, Germany.

In particular, I would like to express my appreciation and sincere gratitude towards my supervisor Prof. Dr.-Ing. Mario Pacas, the Head of the Institute. He gave me the opportunity of carrying out this work, whose completion would have not been possible without his support and guidance throughout all its stages.

I would also like to thank Prof. Dr.-Ing. Friedrich W. Fuchs of the Christian-Albrechts University of Kiel for his work as co-examiner. The corrections and comments he proposed helped to improve the present manuscript.

I wish to express my gratitude towards Prof. Dr. rer. nat. Rainer Patsch for his guidance of the doctoral procedure.

I am obliged to all the staff members of the Institute for creating an excellent working atmosphere. I had the chance to benefit from them in both a professional and in an interpersonal way.

Last but not least, I want to thank my family for their understanding and constant encouragement.

This work received financial support from the German Research Foundation (Deutsche Forschungsgemeinschaft)

Eva Teresa Serna Calvo

Contents

Nomenclature

Symbols

B	Magnetic flux density
c	Constant
C	Fault indicator
D	Average diameter of the air gap
e(t)	Back-emf
f	Frequency
F	Magnetomotive force
g	Winding function
G(s)	Transfer function
G(jω)	Frequency response
H	Magnetic field intensity
i(t)	Current
$i_{2,k}(t)$	Current in rotor loop number k
$i_{\mu 2}(t)$	Magnetizing current
Im	Imaginary value
J	Rotor inertia
j	Imaginary unit
k	Frequency index
K_P	Proportional controller gain
L	Inductance
L_{ab}	Mutual inductance

L_{aa}, L_{bb}	Self inductance
L_{1h}	Mutual inductance
ℓ	Active core length
m_s	Number of stator windings
$M_i(t)$	Internal or electromagnetic torque
$M_L(t)$	Load torque
N	Number of samples
N_r	Number of rotor bars
p	Pole pair number
q	Cross section
s	Slip/ Laplace operator
R	Resistance
Re	Real value
t	Time variable
T	Time constant
T_I	Integration time constant
T_s	Sampling period
U_{DC}	DC-link voltage
V	Volume
W	Energy
w_s	Number of stator turns
$x(n)$	Discrete-time sequence
X	Fourier transform of x

z	z-domain variable
α	Polar coordinate
α_{Nr}	Electrical angle of the slot of the rotor
β	Pole of the observer
Δ	Increment
γ	Electrical rotor angle
δ	Geometric air gap
$\delta"$	Equivalent air gap
θ	Integration variable for the fault indicator
κ	Electrical conductivity
μ_0	Permeability
ν	Harmonic order
ξ	Winding factor
ρ	Electrical skew angle
σ	Leakage factor
τ_p	Pole pitch
φ_2	Rotor flux angle
χ	Skew factor
ψ	Flux linkage
ω	Angular frequency
$\omega_{h,k}$	Frequency of harmonic component k
ω_2	Slip angular frequency
$\omega_M(t)$	Motor electrical angular velocity

$\Omega_M(t)$	Motor mechanical angular velocity

Matrices

[A]	State matrix of the observer
[B]	Input matrix
[C]	Output matrix
[D]	Direct transmission matrix
[E]	Effect of the disturbance vector
$[Q_B]$	Matrix of observability
[x]	State vector
[u]	Input vector/ Voltage vector
[y]	Output vector
[z]	Disturbance vector
[i]	Current vector
$[\psi]$	Flux vector
[R]	Resistance matrix
[L]	Inductance matrix
$[L_{11}]$	Matrix of the self inductances of each stator phase due to the air gap flux
$[S_1]$	Matrix with the inductances due to the leakage flux of the stator
$[L_{12}]$	Matrix with the mutual inductances between the stator coils and the rotor loops
$[L_{22}]$	Matrix of the self inductances of each rotor loop due to the air gap flux
$[S_2]$	Matrix with the inductances due to the leakage flux of the rotor
m,n,r	Index to denote the order of the matrix

Subscripts Indexes

x_1	Stator
x_2	Rotor
x_α, x_β	Components of a space phasor in the stationary coordinate system
x_d, x_q	Components of a space phasor in the field coordinate system
x_{MAX}	Maximal value of the variable x
x_{min}	Minimal value of the variable x
x_{sl}	Corresponding to the sensorless scheme
x_U, x_V, x_W	Phase U, V and W
x_0	Initial value
$\underline{x}$	Complex quantity, space phasor
x_z	Corresponding to the disturbance
x_{ring}	Corresponding to the portion of ring between two bars
x_{bar}	Corresponding to a bar

Superscripts Indexes

x^*	Reference value/Conjugate
$\hat{x}$	Observed value
$\tilde{x}$	Observer error
$\dot{x}$	Derivative
$[X]^{-1}$	Matrix inverse
$[X]^T$	Matrix transpose
x^{1bb}	Corresponds to the case of one broken rotor bar
x'	Referred to the stator

Abbreviations

AC	Alternating current
CM	Condition monitoring
DC	Direct current
DSP	Digital Signal Processor
DFT	Discrete Fourier Transform
DTFT	Discrete-time Fourier Transform
EMF	Electromotive force
FIR	Finite Impulse Response
MMF	Magnetomotive force
FFT	Fast Fourier Transform
IIR	Infinite Impulse Response
IM	Induction Machine
MCSA	Motor Current Signature Analysis
SDFT	Sliding DFT

1 Introduction

Induction machines (IMs) spread over industrial production lines because of their simple and robust construction, though they do not always operate to their maximum designed life. Several surveys identify as major causes of damage bearing and stator and cage-rotor related faults [1], [2]. Among the rotor faults, there is a significant number of failures associated with cracked or broken bars, especially for those IMs operating under arduous duty cycles [3].

Rotor bars are designed to cope with a certain amount of aging stresses, mainly thermal, electrical, mechanical, and the so called environmental stresses which deal with the type of atmosphere in which the machine operates. During steady-state operation, if IM rating values are not exceeded, maximal useful life is guaranteed provided that installation conditions and maintenance are correct. However, during transient operation as direct-line starting, the IM draws currents higher than the nominal ones for almost the full acceleration time. Moreover, heavy duty cycles like reversing under load or operation under repetitive duty cycles cause overheating in the rotor cage. In addition, small rotor asymmetries can appear as a consequence of manufacture defects, especially in the process of aluminum die-casting.

Broken copper bars in fabricated rotors and blowholes in aluminum die-cast rotors lead to the appearance of characteristic harmonics in different variables of the IM, namely flux, stator current and shaft torque. Thus, the performance of the machine is reduced mainly due to the undesired low frequency mechanical oscillations in the shaft. In addition, a higher current flows through the bars near to the affected bar, causing a temperature rise and an increase in power loss, thus reducing the efficiency of the machine. The evolution of these failures, although a slow process, contributes to a further worsening of the rotor cage. Indeed, this type of fault, if not repaired, can cause secondary effects such as damage to the rotor laminations and even to the stator core and windings.

For IMs connected to the mains, detection of rotor asymmetries at an early stage and the subsequent development of suitable condition monitoring (CM) techniques have been thoroughly investigated in recent decades. With the help of condition-based maintenance procedures, unsafe operation of the IM can be avoided and a higher reliability is achieved than if just conventional maintenance procedures are employed as failure-based maintenance and planned or preventive maintenance [4]. Moreover, CM techniques incur lower economic losses – particularly in machines operating under the stringent demands of availability whose collapse can cause the shutdown of a whole production line.

A growing number of IMs are supplied by inverters which provide voltage of variable amplitude and frequency in order to continuously control their velocity within a certain range. In this case, the existing diagnostic techniques for line-connected machines are not applicable, at least directly, mainly because a constant fundamental frequency of operation can not be assumed. Furthermore, for those applications requiring a fast response of the drive to changes in velocity and torque, closed-loop control techniques as field oriented control are used and then controllers can even compensate in some way for the effects caused by the fault [5], [6]. Despite these issues, the use of modern electrical drives favours the implementation of online diagnostic techniques, since they include current sensors and fast digital processors for the control, which can also be used for diagnostic purposes. For these reasons, a better understanding of the rotor fault mechanism in inverter-fed variable speed IMs, particularly with closed-loop control, is required. Then, suitable diagnostic systems which basically make use of the available processing capabilities and sensors of the drive can be developed.

1.1 State of the art

In the following, the state of the art is divided into three main areas. First area deals with the modeling of IMs with an asymmetric rotor. The other two areas review existing CM techniques for the diagnostics of rotor asymmetries of IMs supplied from the mains and of IMs used in drives equipped with field oriented control.

1.1.1 Generalities

The main objective of the fault detection and diagnostics research areas, widely addressed from several points of views since the early eighties, is to study methodologies for identifying and precisely characterizing possible faults arising in a plant. Terminology in this field may not be consistent in the literature and thus a brief list with the basic terms and definitions used in this work is first introduced. Next, a classification of fault detection techniques is given, which includes a description of their main features.

1.1.1.1 Glossary

- **Availability** is the probability that a system or equipment will operate satisfactorily and effectively at any point of time measurement.

- **Analytical redundancy** is the use of two or more (not necessarily similar) ways to determine a variable, where one way uses an analytical mathematical process model.

- A **constraint** is a limitation imposed by nature (physical laws) or man. It permits the variables to take certain values in the variable space.

- A **disturbance** is an unknown (and uncontrolled) input acting on a system.

- An **error** is a deviation between a measured or computed value (of an output variable) and the true, specified or theoretically correct value.

- A **failure** is the permanent interruption of a system or subsystem ability to perform its required function under specified operating conditions.

- **Failure modes** are the various ways in which a failure occurs.

- **False alarm** is the event that an alarm is generated even though no faults are present.

- A **fault** is an unpermitted deviation of at least one characteristic property of the system from the acceptable or standard condition.

- **Fault accommodation** is to reconfigure the system so that the operation can be maintained in spite of a present fault.

- A **fault effect** is the consequence of a fault on the signals, operation, function, or status of an item.

- **Fault detection** is the determination of faults present in a system and time of detection.

- **Fault diagnosis** is the determination of kind, size, location of a fault and time of appearance, including fault isolation and identification.

- **Fault identification** is the determination of the size and time variant behavior of a fault. Follows fault isolation.

- **Fault isolation** is the determination of the nature, location and time of detection of a fault by evaluating the symptoms. Follows fault detection.

- A **fault-tolerant system** is a system where a fault is accommodated with or without performance degradation, but a single fault does not develop into a failure.

- **Hardware redundancy** is the use of more than one independent instrument to accomplish a given function.

- **Monitoring** is a continuous real time task of determining the conditions of a physical system by recording information, recognizing and indicating anomalies of behavior.

- **Protection** includes those means by which potentially dangerous behavior of the system is suppressed.

- A **qualitative model** of a system describes its behavior with relations among system variables and parameters in heuristic terms, such as causalities or if-then rules.

- A **quantitative model** of a system describes its behavior with relations among system variables and parameters in analytical terms, such as differential or difference equations.

- A **reconfiguration condition** is the worst-case fault effect that is tolerated before a remedial action is executed that accommodates the fault.

- **Reliability** is the ability of a system to perform a required function under stated conditions, within a given scope, during a given period of time.

- A **residual** is a fault indicator based on deviations between measurements and model equation based calculations.

- **Sensor fusion** is the integration of correlated signals from different sensors (information sources) in a single representation or action.

- **Severity** is a measurement of the seriousness of fault effects using verbal characterization. Severity considers the worst-case damage to equipment, damage to environment, or degradation of a system's operation.

- **Supervision** is the monitoring of a physical system, so that appropriate actions are taken in order to maintain the operation in the case of faults.

- A **threshold** is a limit value of a residual's deviation from zero. If exceeded, a fault is declared to be detected.

1.1.1.2   Classification of fault detection techniques

Fault detection techniques are very diverse in terms of their theoretical basis, advantages and drawbacks. The most common classification arises from the method used to distinguish a normal or healthy operating condition from a potential failure model. Under this criterion, signal-based techniques are mostly used and they are considered the classic approach. Lately, model-based techniques have been arising as a feasible alternative for developing fault detection techniques. The main features of both of these groups, signal-based and model-based techniques are briefly described in the following.

Signal-based techniques rely on measured physical quantities which contain information about the fault to be diagnosed and whose analysis allows distinguishing between acceptable and unacceptable operations, without requiring a model of the real process. The measured quantities can be analyzed in the time domain or in the frequency domain or in a combination of both of them. In order to extract the signal features suitable for the diagnostics, techniques of diverse complexity are available. Among them, basic tools as mean square or root mean square values, approaches as limit and trend checking and spectral analysis methods are most frequently used [7].

Model-based techniques rely on a model of the process to perform the diagnosis. This model is used in parallel to the real process and they both have identical inputs. Then, comparing the model outputs to those of the real system, residuals are generated. A great variety of approaches have been developed for the generation of the residuals as discussed in [8]. Methods based on observers, i.e. dynamic algorithms that estimate the state of the process, and methods based on parameter estimation constitute the two main groups. In addition, a distinction is generally made between analytical-based and knowledge-based methods. The former makes use of quantitative mathematical models and the latter of qualitative models based on the available knowledge of the system.

In general, signal-based techniques have been more extensively tested and require minimal process knowledge, initial data and online computation. On the other hand, if model-based techniques are used, the need for instrumentation is reduced and a better adaptability to changing operating conditions is obtained. However, fault detection is in this case sensitive to modeling errors which can cause false alarms. Actually, no technique fulfills all the requirements and the selection of the most suitable one has to be analyzed for each specific application [9].

1.1.2 Modeling of rotor asymmetries in the squirrel cage IM

A good understanding of the behavior of the IM in the presence of rotor asymmetries is required before designing a reliable CM method. The first analyses, dated more than 50 years ago, were mainly theoretical and developed techniques to describe the influence of end-ring faults [10] and broken bars [11] in IM performance during steady-state conditions. Instruments to provide experimental verification of those theoretical analyses and to allow the indication of rotor asymmetries while the IM is in operation were designed later [12], [13], [16]. Then, new investigations appeared that presented more accurate and general representation of open-circuited bars and/or end-rings for machines, also for steady-state analysis [14], [15]. A common result in these studies was the existence of a sideband component in the stator phase current when a rotor fault occurs while the motor is supplied by a sinusoidal and balanced three-phase power supply. Field analysis to evaluate the resulting magnetic field distribution and mechanical performance in the case of broken bars was presented in [17]. Later, models able to include the effect of rotor asymmetries and to represent transient behavior of the IM started to receive attention. In [20], a model based on the space phasor theory to describe the performance of the IM under the presence of broken bars during transients is presented. In [18], a dynamic model based on the coupled magnetic circuit theory also valid to represent end-ring faults is developed. A more general dynamic model for transient analysis even under stator faults appears in [19].

For the scope of the present work where the control system is to be simulated, analytical models are preferred. As to be shown in the followings chapters, a good compromise solution between accuracy and computational efficiency is obtained with the model presented in 2.2.1.

Furthermore, it can be extended to include other asymmetries such as for example stator faults due to some kind of short-circuit of the stator coils [21]. However, it has to be noted that, if only the fundamental wave of the IM is to be considered, suitable coordinate transformations exist that allow reducing this model to an approximate equivalent two-axis one valid for the case of an asymmetrical machine, as demonstrated in [22] for an IM with rotor asymmetries. Though, despite the huge diminution in the order of the resulting system of differential equations, the accuracy of this approximated two-axis model varies with the number of rotor bars and thus it has not been used in this work. Other drawbacks of the two-axis model are that the rotor currents at each bar are not directly calculated and it presents no flexibility to simulate different configurations or the degree of severity of the faults.

1.1.3 CM of rotor asymmetries in the IM with balanced AC voltage supply

Diagnostics of rotor asymmetries in line-connected IMs has been largely investigated in recent decades. Among the abundant literature on the subject, main methods are briefly described in the following.

Air gap search coils. The work on air gap search coils has been applied to smaller machines, especially for measuring torque rather than for diagnostics purposes. Its effectiveness in detecting all kind of winding faults was demonstrated in [17], [24]. However, due to the inconvenience of installing search coils in commercial machines, this technique has not been widely used on motors. Instead, the possibility of using the stator winding itself as a search coil has recently been proposed [26]. The main advantage is that this method enables rotor asymmetry and external mechanical influences to be distinguished and it is simple in terms of signal processing.

Vibration monitoring. Vibration monitoring is a classic tool for CM of rotational machines in general and IM in particular when dealing with faults of mechanical origin as eccentricities and bearings faults [23], [25]. Due to the additional instrumentation required and the high grade of expertise needed to carry on a proper diagnostics, other CMs methods are in general preferred for IM [27]. Recent investigations exist however where vibration monitoring is used in combination with other CM in order to improve the overall effectiveness of the diagnostics [28].

Current monitoring. Monitoring methods based on the motor current are becoming of significant interest. Their main features are that they constitute a non-invasive method and they do not require any additional instrumentation such as vibration sensors. Among the existing CM schemes that rely on the analysis of the motor current, a great amount of research is oriented towards application of Motor Current Signature Analysis (MCSA). The MCSA technique consists basically in the application of Fourier analysis to the stator currents to extract the characteristic harmonics caused by the fault [29]. Thus, this technique is applied to IMs operating under steady-state conditions. In particular, rotor asymmetries cause the appearance of characteristic harmonics with the frequencies $(1\pm2s)f_1$, whose suitability for the diagnostics has been extensively investigated [30], [31], [32]. An alternative approach to extract fault information from the current uses the two-dimensional representation of the space phasor of the stator current for the detection of broken bars and it is known as the Park's Vector Approach [33]. The negative impact of certain time varying loads on the diagnostics by means of current monitoring has also been analyzed [34] and several solutions exist so far [36]. Moreover, new techniques to detect broken rotor bars that perform the analysis of the stator current during transients as for example the start-up of the IM have been proposed. Most suitable diagnostic methods based on transient analysis use the wavelet transform [36]. A comparison of signal processing based techniques for the detection of broken bars is presented in [37], which includes both conventional steady-state techniques as well as the approaches based on time-frequency transformations.

Axial leakage flux. Shaft or axial leakage flux of an IM pertains to one of the oldest and classic measurement methods in electrical machines [12]. Rotor asymmetries, as a single broken rotor bar, result in a change in the flux in the axial direction [23], [25]. This change is visible in the spectral analysis of the signal obtained from a search coil wound concentrically with the shaft of the machine, and in particular the amplitude of that voltage can be used as an indicator of any fault in the rotor [18]. This method is also non-invasive and it allows detecting a variety of faults types. However, it is considered complex if compared with MCSA and it is less popular.

Vienna Monitoring Method. This method, developed at the Technical University of Vienna, allows the diagnosis of rotor asymmetries with the help of measured voltages, currents and angular rotor position. The torque is calculated from two different models. For an ideally symmetric machine, torque values from these two models should be equal. Rotor faults, however, lead to different torque values [38]. Its main shortcomings are that some effects like sensor asymmetries or voltage imbalance are not considered and also the parameters of the models need some adaptation techniques to deal with the effect of temperature and saturation. As a result, quantification of the fault severity may be difficult with this method.

AI-based methods. For the sake of completeness, those approaches based on techniques of Artificial Intelligence and knowledge-based systems are also to be mentioned [39].

1.1.4 CM of rotor asymmetries in the IM in inverter closed-loop operation

Closed-loop control is widely used in high performance drives. However, the majority of existing diagnostic techniques deal only with line-fed machines and are insufficient for the fault detection in closed-loop inverter-fed IMs. Moreover, although several approaches for the diagnostics of rotor asymmetries have been proposed in this case, a unique method is difficult to obtain because results are highly influenced by the type of control being used. A review of existing techniques for the diagnostics of rotor asymmetries for drives equipped with field oriented control is presented in the following.

Vienna Monitoring Method. This method which was introduced in the previous subchapter 1.1.3 is mainly adopted in field oriented control drives. Although it delivers good results in both stationary and dynamic cases, voltage sensors are needed and depending on the specifications, a rotor position sensor as well [40], [41].

Pendulum rotary field. The basis of this technique for detecting broken rotor bars is to monitor the orientation of the axis of the IM's air gap flux and hence compute the range of its oscillation in a synchronously rotating coordinate system, which is used as a diagnostic index [42]. This is justified as follows. Under ideal conditions, the magnetic fields of an IM rotate at synchronous speed. However, any asymmetry in the rotor bars disturbs the air-gap magnetic field causing it to oscillate around its original synchronously rotating axis. Although this oscillation may exist even for a healthy machine due to the machine's structural imperfections, this oscillation is significant and detectable in the presence of a single broken bar and progressively increases with the increase in the number of broken bars. The air gap flux is identified from the measured voltages and currents of the machine. Thus, similar to the "Vienna Monitoring", this method also needs additional voltage sensors.

Current monitoring. In the case of controlled electric drives, no general solution exists for the diagnostics of rotor asymmetries merely based on the current monitoring. The possibility of extracting information about the fault based on the Fourier analysis of the components of the stator current space phasor in field coordinates is exploited in [6] for a field oriented controlled IM. The analysis, only valid to stationary conditions, shows that MCSA cannot be directly applied for the diagnostics mainly due to the influence of the tuning of the controllers.

1.2 Goals of this work

This research focuses on the diagnostics of rotor asymmetries in drives equipped with field oriented control. Most of the existing diagnostic methods operate in open-loop. In inverter closed-loop operation, these methods can not directly be applied and thus further work is needed. Significant challenges are added to the fault detection in this case due to the operation at variable frequency, the wide range of operation and the dynamics of the controllers, which try to eliminate the disturbance introduced by the fault. The specific goals of this work are:

a) Derivation of an apropriate IM model for the analysis of the effects of rotor asymmetries

b) Selection of the suitable variable for fault detection in inverter closed-loop operation with field oriented control

c) Development of a computational efficient algorithm for fault detection

d) Addition of CM to electric drive making use of existing sensors and hardware

e) Evaluation of the performance of the method for several variants of field oriented control including encoderless control schemes

f) Validation of the novel diagnostic method through experimental results

1.3 Outline of the chapters

This work is organized as follows: After an introduction and a review of previous works in CM of IMs in Chapter 1, Chapter 2 gives a general dynamic model of an IM based on the coupled magnetic circuit theory which contains a full description of the squirrel cage rotor. This model is modified to include rotor asymmetries, so that the performance of the IM under both healthy and faulty conditions can be described. Also, the well known dynamic equations of the IM based on the space phasor theory are formulated and three different variants of the field oriented control are examined.

In Chapter 3, the behavior of an IM under the presence of rotor asymmetries is discussed. In particular, the characteristic frequencies caused by those asymmetries, firstly in the case of balanced three-phase AC voltage supply, and secondly in the case of inverter closed-loop operation in steady-state conditions are analyzed. In this second case, not only the terminal variables, but also the control ones are included in the analysis and an evaluation of the most suitable variable for the diagnostics is made.

In Chapter 4 a new method is proposed to detect rotor asymmetries in the case of field oriented controlled IMs. This new method is able to identify online the characteristic component introduced by the rotor asymmetry based on the Fourier series. After describing the main algorithms for the digital implementation of Fourier analysis, the most suitable one in terms of less demanding computational efforts is selected. Also its validity in dynamic operation is discussed.

In Chapter 5, the theoretical principles investigated in this work and the proposed method for the diagnostics of rotor asymmetries in field oriented controlled IMs have been experimentally validated.

Finally, Chapter 6 presents the conclusions of this work.

2   Mathematical description of an induction machine

2.1   Set of equations of a general rotating electrical machine

Mathematical models of electrical machines are mostly based on the coupled magnetic circuit theory where the machine is considered as a set of magnetically linked coils. In general, the electromagnetic interactions of an electrical machine without commutator can be described employing the matrix notation by the following set of non linear differential equations

$$[u] = [R][i] + \frac{d}{dt}[\psi] \tag{2.1}$$

with

[u] voltage vector
[i] current vector
[ψ] flux vector
[R] resistance matrix.

In order to fully describe the general AC machine, the mechanical equation is needed which gives the acceleration torque as a result of the torque balance in the shaft of the machine as follows:

$$J \cdot \frac{\ddot{\gamma}}{p} = M_i - M_L, \tag{2.2}$$

where J is the axial moment of inertia of the whole drive, M_i is the internal torque or electromagnetic torque developed by the machine also called "air gap torque" and M_L is the load torque. γ is the later defined electrical angular position of the rotor, which is obtained from multiplying the number of pole pairs p by the mechanical angular position and then

$$\gamma = \gamma_{el} = p \cdot \gamma_{mech}. \tag{2.3}$$

First, the assumptions that have been considered in order to reduce computation time will be given. Analyses take into account only the fundamental wave of the air gap flux and also assume:

- m_s stator windings sinusoidally distributed
- Uniform air gap δ
- N_r evenly distributed bars over the surface of the squirrel rotor cage, which are insulated from the rotor core
- Negligible effects of iron saturation
- Hysteresis, eddy currents and temperature influence in the parameters of the system are neglected
- Friction and windage losses are neglected and the mechanical coupling is considered rigid

Under those assumptions, the flux vector $[\psi]$ can be seen as the product of the inductance matrix $[L]$ with the current vector $[i]$

$$[\psi] = [L][i],\qquad(2.4)$$

where the resulting flux vector consists of the individual coil fluxes which in the general case are dependent on all the currents flowing in the machine.

The resistance matrix $[R]$ contains only constant elements and the inductance matrix $[L]$ includes both constant inductances as the self and leakage flux inductances and time variant inductances as the mutual ones which depend on the relative angular position between stator and rotor.

The internal torque developed by the rotating machine is deduced in [43] and is given by

$$M_i = \frac{1}{2}p[i]^T\left\{\frac{d}{d\gamma}[L]\right\}[i].\qquad(2.5)$$

For the purpose of digital simulation, dynamic equations (2.1) to (2.4) are expressed in state-space form as follows:

$$\frac{d}{dt}[i] = [L]^{-1}\left([u]-[R][i]-\left\{\frac{d}{d\gamma}[L]\right\}[i]\dot\gamma\right)$$
$$\frac{d}{dt}\dot\gamma = \frac{p}{J}\left(\frac{p}{2}[i]^T\left\{\frac{d}{d\gamma}[L]\right\}[i]-M_L\right)\qquad(2.6)$$
$$\frac{d}{dt}\gamma = \dot\gamma$$

As can be seen, the currents, the electrical angle and the velocity are selected as the state variables, whereas the voltages and the load torque M_L are the selected input variables. To complete the model, the elements of matrices [R] and [L] have to be calculated. Next, the analytical expressions which define them will be deduced for the case of a squirrel cage IM.

2.2 Determination of the inductances due to the air gap flux

The inductances due to the air gap flux are obtained based on the approach presented in [42]. This approach assumes no symmetry in the placement of any motor coils in the slots and thus can be extended to the case of asymmetric rotors, as it will be explained in subchapter 2.4. Next, only the important steps for the derivation of the inductances will be presented. It consists of calculating the stored magnetic energy in the air gap of a machine from the volume integral of the magnetic energy density

$$W_m = \frac{1}{2} \iiint B \cdot H \cdot dV = \frac{1}{2} \mu_0 \iiint H^2 \cdot dV, \tag{2.7}$$

which under the assumption of linear magnetic circuit can also be expressed as

$$W_m = \frac{1}{2} [i]^T [L][i]. \tag{2.8}$$

Then, the elements of the matrix [L] are obtained from the simple comparison term to term of both expressions (2.7) and (2.8).

If a two-pole machine consisting of two diametral coils as shown in Fig. 2.1 is considered, then the volume element is equal to

$$dV = \frac{1}{\pi} \cdot \ell \cdot \tau_p \cdot \delta''(\alpha, t) \cdot d\alpha \tag{2.9}$$

where α is the polar coordinate, ℓ is the active length of the machine, $\delta''(\alpha, t)$ is the equivalent air gap which is enlarged in comparison to the geometrical air gap δ to include the influence of slots in stator and rotor [44] and τ_p is the pole pitch which is defined as

$$\tau_p = \frac{\pi \cdot D}{2 \cdot p} \tag{2.10}$$

with D equal to the average diameter of the air gap (here p= 1).

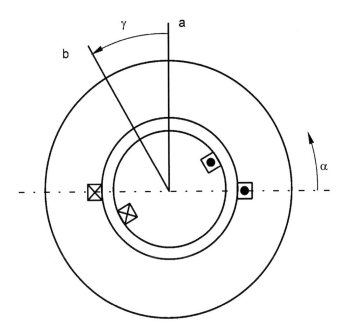

Fig. 2.1   Model of the machine with two diametral coils.

With help of equations (2.9) and (2.10), the volume integral (2.7) yields

$$W_m = \frac{1}{2\pi} \cdot \mu_0 \cdot p \cdot \ell \cdot \tau_p \cdot \int_0^{2\pi} H^2(\alpha, t) \cdot \delta''(\alpha, t) \cdot d\alpha \qquad (2.11)$$

where $H(\alpha, t)$ represents the average radial component of the magnetic field in the air gap.

The resulting magnetomotive force (MMF) if the Ampere's circuit law is applied equals

$$F(\alpha, t) = H(\alpha, t) \cdot \delta''(\alpha, t). \qquad (2.12)$$

On the other hand, the MMF can also be obtained from multiplying the instantaneous current flowing through the winding by a function represented by $g(\alpha, t)$, which describes the geometrical dimensions of the winding as well as its spatial position in the stator laminated core

$$F(\alpha, t) = g(\alpha, t) \cdot i(t). \qquad (2.13)$$

In the following, g will be called winding function.

For the simplified machine depicted in Fig. 2.1, the resulting stored magnetic energy from substituting (2.12) and (2.13) in (2.11) yields

$$W_m = \frac{1}{2\pi} \cdot \mu_0 \cdot p \cdot \ell \cdot \tau_p \cdot \int_0^{2\pi} \frac{1}{\delta''(\alpha)} \cdot \left(g_a^2 \cdot i_a^2 + 2 \cdot g_a \cdot g_b \cdot i_a \cdot i_b + g_b^2 \cdot i_b^2 \right) \cdot d\alpha \qquad (2.14)$$

and equation (2.8) yields

$$W_m = \frac{1}{2} \cdot L_{aa} \cdot i_a^2 + L_{ab} \cdot i_a \cdot i_b + \frac{1}{2} \cdot L_{bb} \cdot i_b^2. \qquad (2.15)$$

Then, the expression for the mutual inductance between the two coils in Fig. 2.1 is equal to:

$$L_{ab} = \frac{1}{\pi} \cdot \mu_0 \cdot p \cdot \ell \cdot \tau_p \cdot \int_{\alpha=0}^{2\pi} \frac{1}{\delta''(\alpha)} \cdot g_a(\alpha, t) \cdot g_b(\alpha, t) \cdot d\alpha \qquad (2.16)$$

and the self inductance for the stator coil L_{aa} is equal to

$$L_{aa} = \frac{1}{\pi} \cdot \mu_0 \cdot p \cdot \ell \cdot \tau_p \cdot \int_{\alpha=0}^{2\pi} \frac{1}{\delta''(\alpha)} \cdot g_a^2(\alpha, t) \cdot d\alpha . \qquad (2.17)$$

Finally, L_{bb} is calculated in a similar way for the rotor coil. Winding functions depend on the polar coordinate α and mutual inductances depend on the later defined electrical angular position γ. A standard way to simplify the calculations considers the representation of the winding functions with their corresponding equivalent Fourier series [43], as will be explained in the next subchapter for the case of an IM with squirrel cage rotor.

2.3   Model of the squirrel cage induction machine

The general set of equations of the IM considering the complete description of the squirrel cage is here determined in the natural reference frame, i.e. referring the quantities of stator and rotor to their respective coordinate systems. A main drawback of the developed model is that the matrix of mutual inductances between the stator coils and the rotor loops depends on the angular position. However, this model facilitates the incorporation of rotor asymmetries and thus is capable of describing the IM response in transient as well as in steady-state modes of operation under faulty conditions.

Fig. 2.2 shows a simplified scheme of the resulting model corresponding to a squirrel cage IM with $m_s = 3$ stator windings and $N_r = 28$ rotor bars. The stator is represented by three identical coils shifted $\frac{2\pi}{3}$ electrical radians in the space and the rotor cage is also considered symmetrical and modeled in terms of current loops, where each rotor loop comprises two adjacent rotor bars and those portions of end-ring which join. As stated in the previous subchapter, the flux can be eliminated in (2.1) with the help of expression (2.4), so that the resulting voltage equations of the machine are equal to

$$\begin{bmatrix}[u_1]\\[u_2]\end{bmatrix}_{n\times1} = \begin{bmatrix}[R_1] & [0]\\[0] & [R_2]\end{bmatrix}_{n\times n} \cdot \begin{bmatrix}[i_1]\\[i_2]\end{bmatrix}_{n\times1} + \frac{d}{dt}\left(\begin{bmatrix}[L_{11}]+[S_1] & [L_{12}]\\[L_{12}]^T & [L_{22}]+[S_2]\end{bmatrix}_{n\times n} \cdot \begin{bmatrix}[i_1]\\[i_2]\end{bmatrix}_{n\times1}\right) \quad (2.18)$$

Note that for the sake of abbreviation "$N_r + m_s$" is denoted as "n".

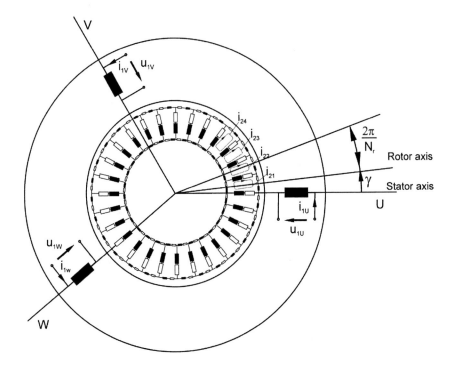

Fig. 2.2   Two-pole symmetrical IM with squirrel cage rotor with $m_s = 3$ and $N_r = 28$.

Next, with expressions (2.2) and (2.5) the mechanical equation results equal to

$$J \cdot \frac{\ddot{\gamma}}{p} = \frac{p}{2} \cdot \begin{bmatrix} [i_1] \\ [i_2] \end{bmatrix}^T_{1 \times n} \cdot \left\{ \frac{d}{d\gamma} \begin{bmatrix} [L_{11}]+[S_1] & [L_{12}] \\ [L_{12}]^T & [L_{22}]+[S_2] \end{bmatrix}_{n \times n} \right\} \cdot \begin{bmatrix} [i_1] \\ [i_2] \end{bmatrix}_{n \times 1} - M_L \qquad (2.19)$$

where γ is defined as the electrical angle between the stator winding U and the first loop of the rotor, as depicted in Fig. 2.2. Vectors $[u_1]$ and $[i_1]$ contain the voltages and currents per phase of the stator windings and are equal to

$$[u_1]^T = \begin{bmatrix} u_{1U} & u_{1V} & u_{1W} \end{bmatrix} \qquad (2.20)$$

$$[i_1]^T = \begin{bmatrix} i_{1U} & i_{1V} & i_{1W} \end{bmatrix}. \qquad (2.21)$$

Regarding the rotor cage, the rotor voltage vector $[u_2]$ is equal to zero

$$[u_2]^T = \begin{bmatrix} u_{2,1} & u_{2,2} & ... & u_{2,N_r} \end{bmatrix} = \begin{bmatrix} 0 & 0 & ... & 0 \end{bmatrix} \qquad (2.22)$$

and the current vector $[i_2]$ contains the rotor loop currents and is equal to

$$[i_2]^T = \begin{bmatrix} i_{2,1} & i_{2,2} & ... & i_{2,Nr} \end{bmatrix}; \qquad (2.23)$$

where the rotor bar currents $i_{bar,k}$ are obtained from loop currents as follows:

$$i_{bar,k} = i_{2,k+1} - i_{2,k}, \qquad k = 1...N_r \qquad (2.24)$$

for each loop k.

The resistance matrix $[R_1]$ results in a square matrix 3 by 3 with the resistance value of each stator winding in the diagonal. However, the rotor resistance matrix $[R_2]$ is not diagonal, due to the galvanic coupling between the adjacent bars. Instead $[R_2]$ is a cyclic symmetrical matrix of dimension N_r, whose components consist of the bar and end-ring resistances values R_{bar} and R_{ring}, respectively.

Regarding the inductance matrices, $[L_{11}]$ contains the self inductances of each stator phase due to the air gap flux, whereas $[S_1]$ refers to the inductances due to the leakage flux of stator, being both a cyclic symmetric 3 by 3 matrix. Leakage flux is characterized because it completes its path only through one of the windings, in this case the stator. The mutual inductance matrix $[L_{12}]$ is of dimension 3 by N_r and comprises the mutual inductances between the stator coils and the rotor loops which, as explained previously, vary with the angle γ depicted in Fig. 2.2. Dealing with the rotor cage, $[L_{22}]$ and $[S_2]$ contain the inductances due to the air gap field and to the leakage fluxes of the rotor, respectively, being both a cyclic symmetrical N_r by N_r matrix. $[S_2]$ is generally defined in terms of the bar and end-ring resistances values, L_{bar} and L_{ring}, respectively.

Formulation of these matrices is a fundamental problem for determining the mathematical description of the machine, and different approaches can be found in the literature [10], [22], [43]. Next, the analytical evaluation adopted to determine the parameters of the model used in this work is described.

2.3.1 Stator system

The stator is assumed symmetrical in this work and consisting of three identical windings ($m_s = 3$) displaced from each other by $\frac{2\pi}{3}$ electrical radians in space, as shown in Fig. 2.2. Then, the corresponding stator resistance matrix $[R_1]$ is equal to

$$[R_1] = \begin{bmatrix} R_1 & 0 & 0 \\ 0 & R_1 & 0 \\ 0 & 0 & R_1 \end{bmatrix}_{m_s \times m_s} \tag{2.25}$$

where R_1 is the resistance value of each stator winding. The self inductances of the stator due to the main flux are contained in $[L_{11}]$, which is equal to

$$[L_{11}] = \tilde{L}_1 \cdot \begin{bmatrix} 1 & -1/2 & -1/2 \\ -1/2 & 1 & -1/2 \\ -1/2 & -1/2 & 1 \end{bmatrix}_{m_s \times m_s} \tag{2.26}$$

with

$$\tilde{L}_1 = \frac{4 \cdot \mu_0 \cdot \ell \cdot \tau_p}{\pi^2 \cdot \delta''} \cdot \left(w_s \cdot \xi_{11} \right)^2 \tag{2.27}$$

where $(w_s \cdot \xi_{11})$ is the number of effective turns of each winding, being w_s the number of turns of each winding and ξ_{11} the so called winding factor [43].

The diagonal elements in (2.26) correspond to the self inductances of each winding phase, whereas the remaining elements correspond to the coupling inductances of each individual winding with the others. Regarding the leakage flux in the stator, $[S_1]$ contains the inductances originated by the leakage flux in the slots and the end windings and is equal to

$$[S_1] = \begin{bmatrix} S_{1ii} & S_{1ik} & S_{1ik} \\ S_{1ik} & S_{1ii} & S_{1ik} \\ S_{1ik} & S_{1ik} & S_{1ii} \end{bmatrix}_{m_s \times m_s} \tag{2.28}$$

The calculation of these inductances is traditionally made in terms of the stored magnetic energy in the slot or end-ring and thus the corresponding geometry and magnetic permeance are needed. Since the exact analytical determination is difficult, empirical correction factors are generally accepted to its approximate calculation [45].

2.3.2 Rotor system

The rotor is considered to comprise N_r identical bars uniformly distributed over the rotor core with a shift equal to $\frac{2\pi}{N_r}$ between each other. Each bar and each end-ring element are described by certain impedance as it is illustrated in Fig. 2.3, which are R_{bar} and L_{bar} for the rotor bar and R_{ring} and L_{ring} for the end-ring. Since this work deals with rotor asymmetries, special attention is given to these parameters as they have to be calculated. In the following, the skin effect has not been considered. It is thus assumed that the distribution of the currents in the bars and end-rings is uniform. Therefore, the calculation of the rotor resistances and inductances does not include any correction coefficient regarding this phenomenon.

The matrix $[R_2]$ represents the rotor resistance and conversely to the stator winding, is not diagonal since there is no galvanic separation between the rotor bars. $[R_2]$ results in a cyclic and symmetric matrix of dimension N_r as follows:

$$[R_2] = \begin{bmatrix} R_0 & -R_{bar} & 0 & \cdots & 0 & -R_{bar} \\ -R_{bar} & & & & & 0 \\ 0 & & \ddots & & & \vdots \\ \vdots & & \ddots & \ddots & & 0 \\ 0 & & & & & -R_{bar} \\ -R_{bar} & 0 & \cdots & 0 & -R_{bar} & R_0 \end{bmatrix}_{N_r \times N_r} \tag{2.29}$$

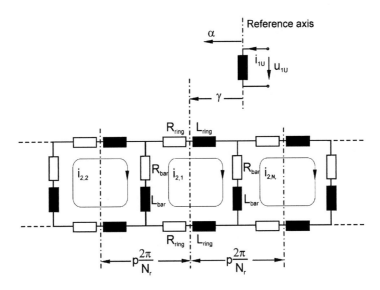

Fig. 2.3   Equivalent circuit of the squirrel cage of an IM where the angular position is referred to the axis of the stator phase U.

with

$$R_0 = 2 \cdot \left(R_{bar} + R_{ring} \right). \tag{2.30}$$

The end-ring resistance R_{ring} is given by

$$R_{ring} = \frac{\pi \cdot D_{ring}}{N_r \cdot q_{ring} \cdot \kappa_{ring}} \tag{2.31}$$

where D_{ring} is the average diameter of the ring, q_{ring} is the cross section of the ring and κ_{ring} the electrical conductivity of the ring material.

Similar to the stator, rotor inductances consist of inductances due to the main flux which are contained in $[L_{22}]$ and inductances due to leakage fluxes in the rotor which are included in $[S_2]$. Starting with the former, two cases are distinguished depending on the consideration of different loops or only one loop of the rotor cage. In both cases, inductances can be obtained from expression (2.16). First, the case with $a{\neq}b$ corresponds to the inductance between two different loops of the rotor cage and it yields

$$L_{2,a\neq b} = \frac{2 \cdot \mu_0 \cdot \ell \cdot \tau_p}{\delta'' \cdot N_r} \cdot \frac{1}{N_r}. \tag{2.32}$$

Secondly, the case with a= b corresponds to the self inductance of one loop and it results

$$L_{2,a=b} = \frac{2 \cdot \mu_0 \cdot \ell \cdot \tau_p}{\delta'' \cdot N_r} \cdot \left(1 - \frac{1}{N_r}\right). \tag{2.33}$$

Then, the matrix $[L_{22}]$ can be build similarly to (2.26) as follows:

$$[L_{22}] = L_{2,k} \cdot \begin{bmatrix} 1 - \dfrac{1}{N_r} & \dfrac{-1}{N_r} & \cdots & & \dfrac{-1}{N_r} \\ \dfrac{-1}{N_r} & & & & \\ \vdots & \ddots & \ddots & \ddots & \vdots \\ & & & \dfrac{-1}{N_r} & \\ \dfrac{-1}{N_r} & \cdots & & \dfrac{-1}{N_r} & 1 - \dfrac{1}{N_r} \end{bmatrix}_{N_r \times N_r} \tag{2.34}$$

with

$$L_{2,k} = \frac{2 \cdot \mu_0 \cdot \ell \cdot \tau_p}{\delta'' \cdot N_r}. \tag{2.35}$$

And the calculation of the leakage flux inductance associated with the bar L_{bar}, is similar to the one corresponding to the stator. Obviously, in this case, the geometry of the rotor slot has to be considered in order to determine the corresponding magnetic permeance. Dealing with the leakage flux associated to the end-ring, the empirical correction given in [45] is here adopted, resulting in the subsequent inductance value:

$$L_{ring} \approx 0.35 \cdot \mu_0 \cdot \frac{\pi \cdot D_{ring}}{N_r}. \tag{2.36}$$

Then, analogously to the case of the rotor resistance matrix, the matrix of leakage rotor flux inductances $[S_2]$ results in a cyclic and symmetric matrix of dimension N_r equal to

$$[S_2] = \begin{bmatrix} L_0 & -L_{bar} & 0 & \cdots & 0 & -L_{bar} \\ -L_{bar} & & & & & 0 \\ 0 & & \ddots & & & \vdots \\ \vdots & & \ddots & \ddots & & 0 \\ 0 & & & & & -L_{bar} \\ -L_{bar} & 0 & \cdots & 0 & -L_{bar} & L_0 \end{bmatrix}_{N_r \times N_r} \tag{2.37}$$

where the diagonal elements are equal to the leakage inductance corresponding to a rotor loop k which is

$$L_0 = 2 \cdot \left(L_{bar} + L_{ring} \right) \tag{2.38}$$

and the remaining elements of $[S_2]$ which represent the leakage flux coupling between bars are equal to $-L_{bar}$ or zero, depending on it corresponds to two adjacent bars or not.

2.3.3 Coupling between stator and rotor

The mutual inductance between a stator winding and a rotor loop is deduced next. First, the corresponding winding functions are given. For the stator winding, the winding function g_{1a} is

$$g_{1a}(\alpha_1, t) = \begin{cases} \dfrac{w_s}{2p} & \text{for } -\dfrac{\pi}{2} < \alpha_1 < \dfrac{\pi}{2} \\ -\dfrac{w_s}{2p} & \text{otherwise} \end{cases} \tag{2.39}$$

where w_s represents the number of stator turns. For the rotor, the winding function g_{2b} is

$$g_{2b}(\alpha_2, t) = \begin{cases} 1 - \dfrac{1}{N_r} & \text{for} -\dfrac{p\pi}{N_r} < \alpha_2 < \dfrac{p\pi}{N_r} \\ -\dfrac{1}{N_r} & \text{otherwise} \end{cases} \tag{2.40}$$

where $\alpha = \alpha_1 = p \cdot \alpha_2$. In Fig. 2.4, the resulting MMF of the rotor loop 2b is depicted. As explained in the introduction, a standard way to simplify the calculations considers the representation of the winding functions with their corresponding equivalent Fourier series [43] and thus the winding functions result equal to

$$g_{1a}(\alpha,t) = \left(\frac{w_s}{2p}\right)\left(\frac{4}{\pi}\sum_{v_1=1}^{\infty}\frac{1}{v_1}\sin\left(\frac{v_1\pi}{2}\right)\cos(v_1\alpha)\right) = \frac{2w_s}{\pi p}\sum_{v_1=1}^{\infty}\frac{1}{v_1}\sin\left(\frac{v_1\pi}{2}\right)\cos(v_1\alpha) \tag{2.41}$$

and

$$g_{2b}(\alpha,t) = \frac{2}{\pi}\left(\sum_{v_2=1}^{\infty}\frac{1}{v_2}\sin\left(\frac{v_2\pi}{N_r}\right)\cos\left(\frac{v_2(\alpha-\gamma)}{p}\right)\right) \tag{2.42}$$

where v_1 and v_2 are the harmonic order for stator and rotor, respectively. With (2.41), (2.42) and (2.16), the mutual inductance between the stator winding 1a and the rotor loop 2b is given by the following expression

$$L_{1a,2b} = \frac{\mu_0\cdot p\cdot \ell\cdot \tau_p}{\pi\cdot\delta''}\cdot\int_{\alpha=0}^{2\pi}g_{1a}(\alpha,t)\cdot g_{2b}(\alpha,t)\cdot d\alpha \tag{2.43}$$

which after some algebraic manipulation results in:

$$L_{1a,2b} = \frac{4\cdot\mu_0\cdot\ell\cdot\tau_p\cdot w_s}{\pi^2\cdot p\cdot\delta''}\cdot\sum_{v=1}^{\infty}\left\{\frac{1}{v^2}\sin\left(\frac{v\pi}{2}\right)\sin\left(\frac{vp\pi}{N_r}\right)\cos(v\gamma)\right\}. \tag{2.44}$$

Therefore the mutual inductance presents harmonics with respect to the electrical angle γ whose influence decreases as the harmonic order v increases. Although some works include the higher space harmonics for the calculation of the mutual inductances [19], in the following only the fundamental component is considered. Therefore, the space distribution of the mutual inductance is assumed to be sinusoidal which analytically implies $v=1$ in (2.44) and it yields

$$L_{1a,2b} \approx \frac{4\cdot\mu_0\cdot\ell\cdot\tau_p\cdot w_s}{\pi^2\cdot p\cdot\delta''}\cdot\sin\left(\frac{p\pi}{N_r}\right)\cos(\gamma) = L_{12,k}\cos(\gamma). \tag{2.45}$$

In the case of study, a= 1,2,3 and b= 1,2,...,N_r and thus $L_{12,k}$ is equal to

$$L_{12,k} \approx \frac{4\cdot\mu_0\cdot\ell\cdot\tau_p}{\pi^2\cdot p\cdot\delta''}\cdot(w_s\cdot\xi_{11})\left(\chi_1\cdot\sin\left(\frac{p\pi}{N_r}\right)\right) \tag{2.46}$$

where χ_1 is the skew factor which is defined as

$$\chi_1 = \frac{\sin\rho}{\rho} \tag{2.47}$$

with ρ equal to the electrical skew angle.

The matrix $[L_{12}]$, whose elements are the coupling inductances between the stator windings and the loops of the rotor, results in

$$[L_{12}] = L_{12,k} \begin{bmatrix} \cos(\gamma) & \cos(\gamma - \alpha_{Nr}) & \cdots & \cos(\gamma - (N_r - 1)\alpha_{Nr}) \\ \cos\left(\gamma + \frac{4\pi}{3}\right) & \cos\left(\gamma + \frac{4\pi}{3} - \alpha_{Nr}\right) & \cdots & \cos\left(\gamma + \frac{4\pi}{3} - (N_r - 1)\alpha_{Nr}\right) \\ \cos\left(\gamma + \frac{2\pi}{3}\right) & \cos\left(\gamma + \frac{2\pi}{3} - \alpha_{Nr}\right) & \cdots & \cos\left(\gamma + \frac{2\pi}{3} - (N_r - 1)\alpha_{Nr}\right) \end{bmatrix}_{3 \times N_r} \tag{2.48}$$

where α_{Nr} is the electrical angle of the slot of the rotor and is equal to

$$\alpha_{Nr} = \frac{2 \cdot p \cdot \pi}{N_r}. \tag{2.49}$$

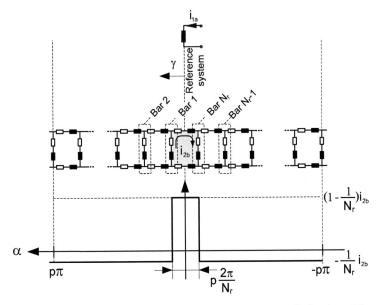

Fig. 2.4   Determination of the mutual inductance between the stator winding 1a and the rotor loop 2b.

Then, the mutual inductance depends on the angle between the phase of the stator winding and the rotor loop being considered, and thus it depends on the electrical rotor position γ, as depicted in Fig. 2.4. The resulting matrix $[L_{12}]$ is neither symmetrical nor cyclic.

2.3.4 Mechanical system and torque calculation

As stated in the assumptions, the coupling of the machine to the load is rigid and thus the mechanical equation corresponds to expression (2.2) where the inertia J and the external load torque M_L are explicitly known. The electromagnetic torque M_i produced by the machine is calculated from expression (2.5) and it can also be expressed in the following simplified form:

$$M_i = p[i_1]_{1\times3}^T \left\{ \frac{d}{d\gamma}[L_{12}]_{3\times N_r} \right\} [i_2]_{N_r\times3} .$$

(2.50)

2.3.5 Set of equations of the squirrel cage IM

As explained in the introduction, for the purpose of digital simulation, electrical and mechanical equations of the IM (2.18) and (2.19) are expressed in state space form. Together with the currents, the electrical angle γ and the angular velocity $\dot{\gamma}$ are the selected state variables. And, given the coefficients of both electrical and mechanical parameters, the needed input data are the load torque and the stator voltages, since for a squirrel cage IM the rotor voltages are zero. Then, considering m_s windings in the stator and N_r bars in the rotor, the following differential set of equations of order (N_r+m_s+2) yields

$$\frac{d}{dt}[i] = [L]_{n\times1}^{-1} \left([u]_{n\times1} - [R]_{n\times n}[i]_{n\times1} - \left\{ \frac{d}{d\gamma}[L]_{n\times n} \right\} [i]_{n\times1} \cdot \dot{\gamma} \right)$$

$$\frac{d}{dt}\dot{\gamma} = \frac{p}{J} \left(\frac{p}{2}[i]_{n\times1}^T \left\{ \frac{d}{d\gamma}[L]_{n\times n} \right\} [i]_{n\times1} - M_L \right)$$

$$\frac{d}{dt}\gamma = \dot{\gamma}$$

(2.51)

where the vector $[i]_{n\times1}$ contains the currents flowing in the m_s stator windings and the currents flowing in the N_r rotor loops. As previously described, each rotor loop comprises two adjacent rotor bars and the end-ring segments that join them (Fig. 2.2). Then, the current flowing through each rotor bar can be calculated from (2.24). Regarding the current circulating in the end-rings, it equals zero provided that all ring segments are intact and similar [14]. In addition, assuming balanced stator winding, no coupling exists between it and the rotor end-ring loop and therefore it does not contribute to the internal torque developed by the IM. As a result, the end-ring current is not considered as state variable which reduces the number of unknown rotor currents by 1.

2.4   Model of the squirrel cage IM with rotor asymmetries

Several analyses of machines with rotor asymmetries and, in particular, broken bars can be found in the literature, as presented in the introduction. In the following, different approaches are analyzed which are based on the model explained in the previous subchapter including the full description of the squirrel cage of the rotor. The condition of broken bar is incorporated in all cases by imposing that no current flows through the affected bar. This condition is accomplished differently by each approach. First, a simplified model is introduced which makes use of the superposition principle and allows justifying in a simple way the effects and causes of the rotor asymmetries in the ideal case of steady-state and sinusoidal voltage supply. Next, two approaches are described which are valid to build up dynamic simulations. One of them represents the affected bar by an open-circuit and the second one increases the resistance of the bar. It will be demonstrated that despite being different in form, the two models are equivalent.

2.4.1 Simplified approach

As explained, if a bar is broken, then no current flows through it. This condition is attained in [15] using the principle of superposition as explained below.

A second set of rotor currents is superimposed on those normally found in a symmetrical machine. This second set is obtained by injecting a current into the affected bar of equal and opposite magnitude to that which would flow in the symmetrical case. Then, under the assumption of a linear magnetic circuit, the rotor of the machine can be represented as the superposition of two configurations, a healthy and a faulty one, as shown in Fig. 2.5.a) and b), respectively. Thus, the effects caused by the rotor asymmetry can be analyzed separately by solving circuit in Fig. 2.5. b). In [15], it is analytically demonstrated that the distribution of the rotor currents flowing in the cage generated by the "faulty current" distributes evenly among the remaining bars or circulates only through the adjacent ones, depending on the slip frequency and the characteristics of the rotor cage.

Although this approach is only valid for steady-state operation and it does not include the effect of the stator in the analysis, it permits a comprehensive and simple analysis of the effects of the rotor asymmetry in the behavior of the machine, as shown in next chapter.

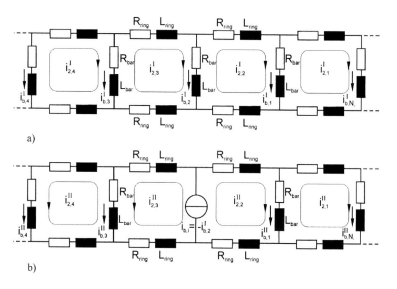

Fig. 2.5   Model of a rotor asymmetry using the principle of superposition [8]: a) equivalent circuit of the healthy rotor, and b) equivalent circuit of the "asymmetric rotor" with the additional injected current $i_{b,i}$.

2.4.2 Open-circuit in place of the broken bar

With this second approach, the broken bar is considered an open-circuit, so that the two loops adjacent to the affected bar are merged into one. Thus, proper relationships between the rotor variables are needed and the inductances of the new loop have to be accordingly recomputed as well. If for example the second bar is considered to be broken, ring currents of loops 2 and 3 are equal and flow in a double width loop. This situation is depicted in Fig. 2.6 where

$i_{2,2}$ and $i_{2,3}$ are replaced by a single current called $i_{2,b'}$. The condition $i_{2,2} = i_{2,3}$ implies that the corresponding columns to $i_{2,2}$ and $i_{2,3}$ are added one to one in the inductance matrix of the rotor $[L_{22}]+[S_2]$. The same relationship is applied to the corresponding rows.

A similar reasoning is applied to the rotor resistance matrix $[R_2]$. In the coupling inductance matrix $[L_{12}]$, the elements corresponding to the affected bar are recalculated. In this case, the rotor is modeled as a mesh with (N_r-1) bars distributed asymmetrically, as it is shown in Fig. 2.7. In particular, if the rotor loop 2b' that includes the broken rotor bar is considered, the winding function is given by:

$$g_{2b'}(\alpha_2, t) = \begin{cases} 1 - \dfrac{2}{N_r} & \text{for } -\dfrac{p\pi}{N_r} < \alpha_2 < \dfrac{3p\pi}{N_r} \\[3mm] -\dfrac{2}{N_r} & \text{otherwise} \end{cases} \tag{2.52}$$

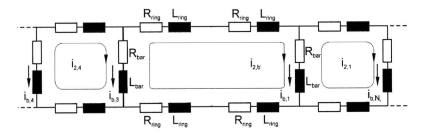

Fig. 2.6   Equivalent circuit of the squirrel cage mesh: open-circuit in place of broken bar.

Operating in a similar way to the case of symmetrical machine, the winding function can be expressed by an expansion in Fourier series:

$$g_{2b'}(\alpha, t) = \frac{2}{\pi} \sum_{v_2=1}^{\infty} \frac{1}{v_2} \sin\left(\frac{v_2 2\pi}{N_r}\right) \cos\left(\frac{v_2(\alpha - \gamma + \gamma_b)}{p}\right) \tag{2.53}$$

where $\gamma_b = \frac{\pi \cdot p}{N_r}$. Then, the mutual inductance between 1a and 2b' by neglecting the harmonics of higher order is approximately equal to:

$$L_{1a,2b'} \approx \frac{4 \cdot \mu_0 \cdot \ell \cdot \tau_p \cdot w_s}{\pi^2 \cdot p \cdot \delta''} \cdot \sin\left(\frac{2 \cdot p \cdot \pi}{N_r}\right) \cdot \cos(\gamma + \gamma_b) = L_{12,k}^{1bb} \cdot \cos(\gamma + \gamma_b) \tag{2.54}$$

with

$$L_{12,k}^{1bb} = c^{1bb} \cdot L_{12,k}, \qquad \text{being } c^{1bb} = \frac{\sin\left(\dfrac{2 \cdot p \cdot \pi}{N_r}\right)}{\sin\left(\dfrac{p \cdot \pi}{N_r}\right)}. \tag{2.55}$$

As shown in Fig. 2.6, due to the single broken bar, two rotor loops are combined in one loop which is denoted as 2b' and thus the mutual inductance $L_{1a,2b'}$ differs from that calculated in (2.45) for the healthy case $L_{1a,2b}$. Finally, the resulting matrix of mutual inductances $[L_{22}^{1bb}]$ is a 3 by (N_r-1) matrix equal to

$$\left[L_{12}^{1bb}\right] = L_{12,k} \begin{bmatrix} \cos(\gamma) & c^{1bb} \cdot \cos\left(\gamma - \dfrac{3}{2}\alpha_{Nr}\right) & \cos(\gamma - 3\alpha_{Nr}) & \cdots \\ \cos\left(\gamma + \dfrac{4\pi}{3}\right) & c^{1bb} \cdot \cos\left(\gamma + \dfrac{4\pi}{3} - \dfrac{3}{2}\alpha_{Nr}\right) & \cos\left(\gamma + \dfrac{4\pi}{3} - 3\alpha_{Nr}\right) & \cdots \\ \cos\left(\gamma + \dfrac{2\pi}{3}\right) & c^{1bb} \cdot \cos\left(\gamma + \dfrac{2\pi}{3} - \dfrac{3}{2}\alpha_{Nr}\right) & \cos\left(\gamma + \dfrac{2\pi}{3} - 3\alpha_{Nr}\right) & \cdots \end{bmatrix}_{3 \times (N_r-1)} \tag{2.56}$$

which corresponds to the case depicted in Fig. 2.6 where it is considered that the second rotor bar is open-circuited.

Next, the elements of matrix $[L_{22}]$ are deduced again based on the expression (2.16). Two cases are distinguished, first the inductance between one healthy and one loop containing the broken bar and second the self inductance of the loop containing the broken bar. The first case yields the following inductance

$$L_{2,a \neq b'} = \frac{2 \cdot \mu_0 \cdot \ell \cdot \tau_p}{\delta'' \cdot N_r} \cdot \frac{1}{N_r} \cdot \left(\frac{2}{N_r}\right). \tag{2.57}$$

For the second case, the inductance is equal to

$$L_{2,a=b=b'} = \frac{2 \cdot \mu_0 \cdot \ell \cdot \tau_p}{\delta'' \cdot N_r} \cdot \frac{1}{N_r} \cdot \left(2 - \frac{2}{N_r}\right). \tag{2.58}$$

Then, the rotor inductance matrix $[L_{22}^{1bb}]$ is given by the following expression:

$$
\left[L_{22}^{1bb}\right] = L_{2,k}
\begin{bmatrix}
1-\dfrac{1}{N_r} & \dfrac{-2}{N_r} & \dfrac{-1}{N_r} & \cdots & & & \dfrac{-1}{N_r} \\[2mm]
\dfrac{-2}{N_r} & 2\cdot\left(1-\dfrac{2}{N_r}\right) & \dfrac{-2}{N_r} & \cdots & & & \dfrac{-2}{N_r} \\[2mm]
\dfrac{-1}{N_r} & \dfrac{-2}{N_r} & 1-\dfrac{1}{N_r} & \dfrac{-1}{N_r} & & & \dfrac{-1}{N_r} \\[2mm]
& & \dfrac{-1}{N_r} & \ddots & \ddots & & \vdots \\[2mm]
\vdots & \vdots & & & \ddots & & \\[2mm]
& & & & & & \dfrac{-1}{N_r} \\[2mm]
\dfrac{-1}{N_r} & \dfrac{-2}{N_r} & \cdots & & & \dfrac{-1}{N_r} & 1-\dfrac{1}{N_r}
\end{bmatrix}_{(N_r-1)\times(N_r-1)}
\tag{2.59}
$$

where $L_{2,k}$ is given by (2.35).

The stator resistance and inductance matrices remain without change and the corresponding matrices for the rotor in the considered case of one broken rotor bar result in (N_r-1) by (N_r-1) square matrices that can be built from simple inspection of Fig. 2.6. The corresponding rotor resistance matrix $[R_2^{1bb}]$ is equal to

$$
\left[R_2^{1bb}\right] =
\begin{bmatrix}
R_0 & -R_{bar} & 0 & \cdots & 0 & -R_{bar} \\
-R_{bar} & 2\cdot(R_0+R_{bar}) & -R_{bar} & 0 & \cdots & 0 \\
0 & -R_{bar} & R_0 & \ddots & & \vdots \\
\vdots & 0 & \ddots & \ddots & & 0 \\
0 & \vdots & & & & -R_{bar} \\
-R_{bar} & 0 & \cdots & 0 & -R_{bar} & R_0
\end{bmatrix}_{(N_r-1)\times(N_r-1)}
\tag{2.60}
$$

and the matrix of leakage inductances of the rotor matrix $[S_2^{1bb}]$ is equal to

$$
\left[S_2^{1bb}\right] =
\begin{bmatrix}
L_0 & -L_{bar} & 0 & \cdots & 0 & -L_{bar} \\
-L_{bar} & 2\cdot(L_0+L_{bar}) & -L_{bar} & 0 & \cdots & 0 \\
0 & -L_{bar} & L_0 & \ddots & & \vdots \\
\vdots & 0 & \ddots & \ddots & & 0 \\
0 & \vdots & & & & -L_{bar} \\
-L_{bar} & 0 & \cdots & 0 & -L_{bar} & L_0
\end{bmatrix}_{(N_r-1)\times(N_r-1)}
\tag{2.61}
$$

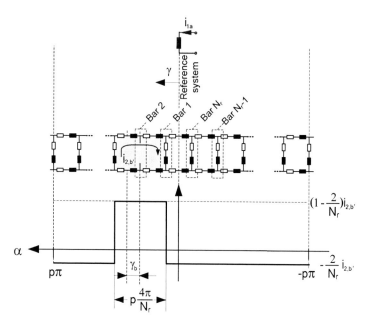

Fig. 2.7   Determination of the mutual inductance between stator winding 1a and rotor loop 2b'. Case of single broken rotor bar.

This is a straightforward strategy that considers the actual topology after the failure. Its implementation is advantageous mainly because of the order reduction of the system to be solved. Nevertheless due to the available computing power this is not a real advantage any more. In addition, the model has to be adapted to each asymmetry.

2.4.3 Increase of the resistance of the broken bar

A broken bar can also be represented as an increase of the resistance of the relevant bar R_{bar} in an amount that depends on the severity of the fault.Then, the order of the system does not change and nor does the required computational effort if compared to the case of healthy machine. Furthermore, this approach is more easily generalized to any number of broken bars, since only the rotor resistance matrix $[R_2]$ has to be recalculated in the model of the machine. To reflect the degree of deterioration of a given bar, a new constant c is introduced so that

$$R_{bar}^{1bb} = c \cdot R_{bar} \tag{2.62}$$

where c=1 means that the bar is intact. Then, the corresponding rotor resistance matrix for the case of a single broken bar placed in the position 2 as depicted in Fig. 2.8 is equal to

$$\left[R_2^{1bb}\right] = \begin{bmatrix} R_0 & -R_{bar} & 0 & \cdots & 0 & -R_{bar} \\ -R_{bar} & (1+c)R_{bar}+2R_{ring} & -c \cdot R_{bar} & & & 0 \\ 0 & -c \cdot R_{bar} & (1+c)R_{bar}+2R_{ring} & -R_{bar} & & \vdots \\ \vdots & & -R_{bar} & R_0 & \ddots & 0 \\ 0 & & & \ddots & \ddots & -R_{bar} \\ -R_{bar} & 0 & \cdots & & 0 & -R_{bar} & R_0 \end{bmatrix}_{N_r \times N_r} \tag{2.63}$$

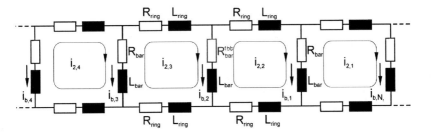

Fig. 2.8   Equivalent circuit of the squirrel cage mesh: increase of broken bar resistance.

This approach is equivalent to the one that assumes the bar to be an open-circuit, for a sufficient large value of resistance R_{bar}^{1bb}, i.e. of c in (2.63). An important practical issue for carrying out the simulations is to additionally increase the inductance of the same bar L_{bar}. This solves numerical problems that may arise during the simulation since the ratio L_{bar}/R_{bar}, i.e. the equivalent time constant of the bar, does not change.

2.5   Model of the symmetric IM by complex state variables

For the subsequent considerations the model of a symmetric IM without failures is needed and is therefore presented in the following. Assuming that the neutral point of the motor is not connected, the equations of the IM can be transformed in a simpler one. They are well known [47] and are presented in the following in a short form. In the general case of a coordinate system rotating at arbitrary positive velocity ω with respect to the stator, the voltage equations are equal to

$$\underline{u}_1 = R_1 \cdot \underline{i}_1 + j \cdot \omega \cdot \underline{\psi}_1 + \underline{\dot{\psi}}_1 \tag{2.64}$$

$$0 = R'_2 \cdot \underline{i}'_2 + j \cdot (\omega - \dot{\gamma}) \cdot \underline{\psi}'_2 + \underline{\dot{\psi}}'_2 \tag{2.65}$$

and the current-flux algebraic relationships are

$$\underline{\psi}_1 = (L_{1h} + L_{1\sigma}) \cdot \underline{i}_1 + L_{1h} \cdot \underline{i}'_2 \tag{2.66}$$

$$\underline{\psi}'_2 = L_{1h} \cdot \underline{i}_1 + (L_{1h} + L'_{2\sigma}) \cdot \underline{i}'_2 \tag{2.67}$$

where $L_{1\sigma}$ and $L'_{2\sigma}$ are the stator and rotor leakage inductances and L_{1h} equals the mutual inductance. The relationships linking these inductances with those presented in the last sections can be found in the pertinent literature [46].

Regarding the mechanical model, the inner torque can be expressed in different ways, depending on the available quantities [46], [47]

$$M_i = \frac{3}{2} \cdot p \cdot \frac{1}{1+\sigma_2} \cdot \text{Im}\left\{ \underline{\psi}'^*_2 \cdot \underline{i}_1 \right\}. \tag{2.68}$$

The internal torque is substituted in the mechanical equation (2.2), which completes the model of the system. In the following, the mechanical angular velocity Ω_{mech} is defined as a function of the electrical angular velocity $\dot{\gamma}$ are and the number of pairs of poles p, according to (2.3). The so called leakage coefficients are used which are defined as a function of the inductances

$$\sigma_1 = \frac{L_{1\sigma}}{L_{1h}}, \quad \sigma_2 = \frac{L'_{2\sigma}}{L_{1h}}, \quad \sigma = 1 - \frac{1}{(1+\sigma_1)(1+\sigma_2)}. \tag{2.69}$$

2.6 Field oriented control principle

The technique called control with field orientation provides the IM decoupled control for the excitation and machine torque [48], [49]. This is achieved by selecting a coordinate system rotating in synchronism with a flux wave (rotor, stator or air gap), which is generally called the field coordinate system.

The principle of field orientation can be explained in a simplified form, if the IM is considered to be supplied by impressed currents, as it would be the case with fast acting current loops. Then, the machine is just characterized by the rotor voltage equation. Assuming the real axis of the field coordinate system to be aligned with the rotor flux space phasor $\underline{\psi}'_2$, the imaginary part of the rotor flux space phasor ψ_{2q} is zero by definition and thus

$$\left| \underline{\psi}'_2 \right| = \psi_{2d} . \tag{2.70}$$

Under this condition, it follows from the voltage equation of the rotor

$$T_2 \cdot \frac{di_{\mu 2}}{dt} + i_{\mu 2} = i_{1d} , \tag{2.71}$$

where $i_{\mu 2}$ is the magnetizing current which is proportional to the amplitude of the rotor flux space phasor ψ_{2d} and $T_2 = \frac{L'_2}{R'_2}$ is the rotor time constant. Eq. (2.71) expresses that the amplitude of the rotor flux space phasor ψ_{2d} can be changed according to a first order delay with the scalar current i_{1d}, which is the projection of the stator current space phasor $\underline{i}_1$ on the real axis of the field coordinate system. Thus, i_{1d} can be defined as the flux producing current component.

On the other hand, the imaginary part of the rotor voltage equation leads to:

$$0 = i_{1q} - \left(\dot{\phi}_2 - \dot{\gamma} \right) \cdot T_2 \cdot i_{\mu 2} . \tag{2.72}$$

Then, it can be deduced that with constant flux magnitude (ψ_{2d} =const.) the scalar current i_{1q}, which corresponds to the imaginary component of the stator current space phasor $\underline{i}_1$ in field coordinates, remains proportional to the rotor slip angular frequency ω_2 equal to

$$\omega_2 = \dot{\phi}_2 - \dot{\gamma} \tag{2.73}$$

where ϕ_2 represents the angle of the rotor flux space phasor $\underline{\psi}'_2$, with respect to the stator and thus it defines the field coordinate system. From (2.68), the inner torque of the machine yields

$$M_i = \frac{3}{2} \cdot p \cdot \frac{1}{1+\sigma_2} \cdot \left| \underline{\psi'_2} \right| \cdot i_{1q} \qquad (2.74)$$

which indicates that if ψ_{2d} is kept constant, the inner torque of the machine M_i can be linearly varied by adjusting i_{1q}. Thus, i_{1q} can be entitled as the torque producing current component. Expressions (2.71) to (2.74), together with the mechanical equation (2.2) constitute a model of the IM in field coordinates, as described by the block diagram in Fig. 2.9.

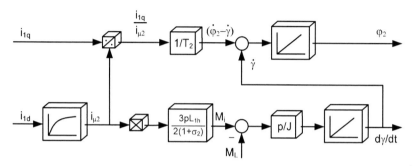

Fig. 2.9   Block diagram of the current-fed IM in field coordinates.

2.6.1   Determination of the rotor flux angle ϕ_2

It has to be noticed that in order to achieve a proper performance, a main requirement of a field oriented control scheme is obtaining an accurate position of the rotor flux space phasor $\underline{\psi'_2}$ which in our case of study is given by angle ϕ_2. Several methods to determine the rotor flux angle ϕ_2 have been developed [46], [50], but it is not the scope of this work to analyze all of them. The main goal is to analyze the performance of the more representative schemes to obtain this information in the presence of rotor asymmetries. As will be seen, the rotor flux angle ϕ_2 can be obtained from currents and angular position or from currents and voltages, based only on the fundamental wave equation of the IM.

2.6.1.1   Current/angle-model

This approach uses the measured stator currents and the rotor position γ. The rotor flux angle ϕ_2 is then obtained by rearranging expression (2.72) as follows

$$\varphi_2 = \int \left(\frac{i_{1q}}{T_2 \cdot i_{1d}} \right) \cdot dt + \gamma. \tag{2.75}$$

The main characteristics of this flux model are that it extends its validity to the whole range of speed. However online-adaptation methods are usually needed to compensate the dependence of the rotor time constant T_2 on temperature and saturation. Also a sensor for the mechanical position of the shaft is necessary.

2.6.1.2 Voltage/current-model

The stator flux space phasor $\underline{\psi}_1$ can be obtained from the stator voltage equation in a stationary reference frame:

$$\underline{u}_1 = R_1 \cdot \underline{i}_1 + \underline{\dot{\psi}}_1 \tag{2.76}$$

and hence

$$\underline{\psi}_1 = \int \left(\underline{u}_1 - R_1 \cdot \underline{i}_1 \right) \cdot dt \ . \tag{2.77}$$

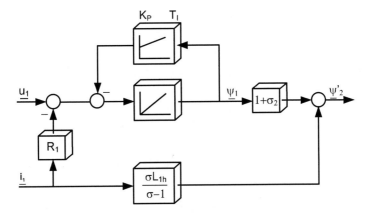

Fig. 2.10   Integration of the stator voltage: approximation with a PI controller [48].

For the rotor flux space phasor $\underline{\psi}'_2$, after some algebraic manipulation of expressions (2.66) and (2.67), follows:

44

$$\underline{\psi'}_2 = \frac{\sigma}{\sigma-1}\cdot L_{1h}\cdot\underline{i_1}+(1+\sigma_2)\cdot\underline{\psi_1}. \tag{2.78}$$

Main advantages of the voltage/current-model are that the stator resistance can be compensated for changes in the temperature. In addition, the influence of saturation in the inductance L_{1h} can be included. Conversely, its main drawback is that the model exhibits a poor performance or it does not work at low frequencies of operation.

For the practical implementation, since the open integration fails at lower frequencies, enhanced integrators can be used [51]. Among others, the integrator can be approximated by a higher order transfer function that can be separated in a pure integrator and a PI controller as feedback, as it is depicted in Fig. 2.10. The transfer function of the structure with the PI controller leads to

$$G(s) = \frac{T_1}{T_1\cdot s^2 + K_p\cdot T_1\cdot s + K_p} \tag{2.79}$$

where K_P is the proportional controller gain and T_1 the integration time constant.

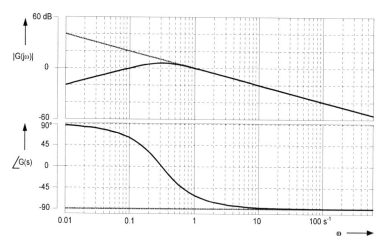

Fig. 2.11   Frequency response of a pure integrator and of an integrator with a PI-controller as feedback.

In Fig. 2.11, the Bode diagram corresponding to (2.79) and of the pure integrator show that the gain at low frequencies of the former is damped to a minimum at low frequencies and even zero as when offsets occur. Up to a few Hertz, gain and phase of both systems correspond, i.e. the new structure behaves like an integrator. It is also a common practice not to use the actual, measured voltages of the inverter; moreover the calculated reference signals with a correction that takes into account the value of the DC-link voltage are used.

2.6.2 Structure of a field oriented control drive

Given the references of the flux producing current i^*_{1d} and the torque producing current i^*_{1q}, the machine can be supplied by means of a current source or a voltage source.

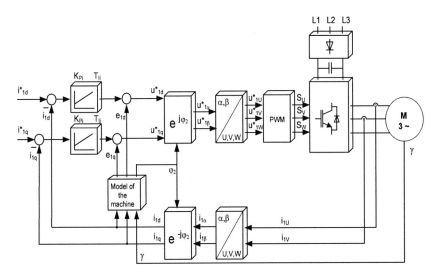

Fig. 2.12   Field oriented control scheme for the IM supplied with voltages.

The structure depicted in Fig. 2.12 is very common since voltage inverters are nowadays the preferable feeding unit for electrical machines. In that case, digital PI controllers are often used to control the previously defined components of the stator current i_{1d} and i_{1q}. With this alternative, the voltages are the quantities to be transformed into the stationary system in order to obtain the corresponding phase voltages at the machine terminals. The variables e_{1q} and e_{1d} represent coupling terms between d- and q- equations that are calculated online and added to the output of the controllers for a decoupling of the plant [52]. Since this alternative is the one used in this work, more details especially dealing with the design of the PI controllers will be presented in the next chapter.

2.6.3 Sensorless velocity control of an IM

The information of the rotor position is commonly provided by position transducers, such as resolvers or optical encoders, which are often undesirable because of space restrictions or the added cost and complexity. State of the art of field oriented control of IMs includes a variety of methods that do not require a transducer to obtain the rotor position γ. Velocity sensorless IM drives are well established in those industrial applications in which persistent operation at lower speed is not considered essential [51]. Indeed, methods based on the so called fundamental wave model have been shown to be able of providing high-performance, field oriented control in the medium to high speed range. In the very low speed range, control algorithms that rely on the tracking of saliencies of the machine using high frequency, carrier signal excitations can provide additional information on the flux angle or the position of the rotor [53]. A brief review of the existing methods is in the following presented.

2.6.3.1 Methods for the estimation of the rotor position on a field oriented controlled IM

A first category of methodologies applied to sensorless velocity control comprises the methods that model the IM by its state equations. A sinusoidal flux density distribution in the air gap is assumed, neglecting space harmonics and other secondary effects. They are either implemented as open-loop structures [53], [54] or as closed-loop observers [55], [56]. The stator model forms the essential component of all model-based approaches. For the open-loop structure, enhanced integrators are needed in the practice as described in 2.6.1.2 and their performance is even though limited to a few Hz. Moreover, as the speed reduces, the sensitivity of the model to a stator resistance mismatch, whose value can vary within a 1:2 range, increases [51]. Then, a model reference adaptive system can be used for parameter identification [54]. Additional problems arise when the stator voltage is not measured and then the method can be improved by considering an exact inverter model. Furthermore, the robustness of open-loop models against parameter mismatch and their dynamics can be improved by using closed-loop observers, that basically use error signals between measured and observed quantities which are fed back to the observer. One example of closed-loop structure to estimate the rotor position is the sliding mode observer [56]. The lower limit of stable operation of fundamental models is reached when the stator frequency is around 1-3 Hz [51].

In the low speed range the anisotropies of the machine are used to identify the rotor position. In essence, transient excitations generated by injected signals having other frequencies than the fundamental or transients caused by inverter switching serve to detect the spatial orientations of existing anisotropies in the machine. Namely, geometrical anisotropy and anisotropy caused by magnetic saturation are distinguished. The geometrical saliency arises from rotor slotting. This effect has been exploited mainly in machines with intrinsic saliencies as the synchronous reluctance machine [52]. In the case of the IM, the effect of the anisotropic structure due to the discrete rotor bars in the resulting current signals to be processed is quite small and difficult to detect [58]. Moreover, they are superimposed on other signals as the load current and distorted because of the nonlinearity of the inverter. On the other hand, the alternative of using a custom designed rotor with an engineered anisotropy is in generally undesirable [59]. Another possibility if no geometrical saliency is present is to exploit the effect of the saturation saliency caused by the fundamental field in the leakage inductances. However, no general methods exist for the estimation of magnetically symmetric IM in the speed region around zero speed [53]. For these reasons, these methods will not be considered in the present work.

2.6.3.2 Structure of a field oriented sensorless velocity control drive

A simplified block diagram of the sensorless control scheme used in this work is depicted in Fig. 2.13, which pertains to the group based on the fundamental wave equation of the IM. In particular, only the voltage/current-model is needed to determine the rotor flux angle φ_2. As described in 2.6.1.2, the stator flux space phasor $\underline{\psi}_1$ is obtained by integrating the stator voltage equation (2.77) which is then used to determine the rotor flux space phasor $\underline{\psi}'_2$. Then, the rotor flux angle φ_2 is obtained from (2.77) and (2.78) as

$$\varphi_2 = \tan^{-1}\left(\frac{\psi'_{2\beta}}{\psi'_{2\alpha}}\right) \tag{2.80}$$

and with (2.75) the position γ of the mechanical angle of the shaft can be found. In a first approach, the simple case that does not compensate the dependence of the rotor time constant T_2 on temperature and saturation is considered (see 2.6.1.1). For the velocity control its actual value is required, which can be obtained by means of the time discrete differentiation of the rotor position information. However, this differentiation amplifies the noise and possible inaccuracies of the position signal. Therefore, an observer is used instead that delivers a better quality of the signal as required for the superimposed velocity control shown in Fig. 2.13. Next, a brief theoretical analysis of this observer is presented.

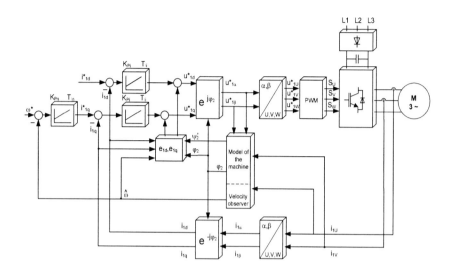

Fig. 2.13   Block diagram of field oriented sensorless velocity control of an IM based on the voltage/current –model for the rotor flux.

2.6.3.3 Velocity observer

As explained, the velocity signal in Fig. 2.13 is obtained by means of an observer instead of a direct differentiation of the rotor angle. In this work, a velocity observer similar to that implemented in [60] has been used. The general form of a state-space model is [61]:

$$\frac{d}{dt}[x]_{nx1} = [A]_{nxn} \cdot [x]_{nx1} + [B]_{nxr} \cdot [u]_{rx1} + [E]_{nxr} \cdot [z]_{rx1}$$

$$[y]_{mx1} = [C]_{mxn} \cdot [x]_{nx1} + [D]_{mxr} \cdot [u]_{rx1}$$

(2.81)

where [x] is the state vector, [u] is the input vector, [y] is the output vector and [z] is the disturbance vector. [A] is the state matrix and [B] is the input matrix. [E] contains the effect of the disturbance vector, [C] is the output matrix and [D] is the direct transmission matrix. The mechanical equation of the machine (2.2) represented in state-space yields:

$$\frac{d}{dt}\begin{bmatrix}\omega\\\gamma\end{bmatrix}=\begin{bmatrix}0&0\\1&0\end{bmatrix}\cdot\begin{bmatrix}\omega\\\gamma\end{bmatrix}+\begin{bmatrix}\dfrac{p}{J}\\0\end{bmatrix}\cdot M_i+\begin{bmatrix}-\dfrac{p}{J}\\0\end{bmatrix}\cdot M_L\left(t\right)$$

$$\gamma=\begin{bmatrix}0&1\end{bmatrix}\cdot\begin{bmatrix}\omega\\\gamma\end{bmatrix}+0\cdot M_i$$

(2.82)

A state observer can be implemented if, and only if, the system is completely observable. This condition is fulfilled if the so called matrix of observability $[Q_B]$ is of rank n or has n linearly independent column vectors. For the case of study, $[Q_B]$ with matrices $[C]$ and $[A]$ from (2.82) yields

$$[Q_B]=\begin{bmatrix}[C]\\[C][A]\end{bmatrix}=\begin{bmatrix}0&1\\1&0\end{bmatrix}$$

(2.83)

The rank of $[Q_B]$ is 2 and thus the system is observable. The disturbance variable M_L is unknown. Still it is possible to consider the disturbance variable if the time characteristic of the disturbance is basically known. In this work, the load torque is assumed to be constant and thus a model of the disturbance can be written as:

$$\frac{d}{dt}[x_s]=[A_s]\cdot[x_s];\ [z]=[C_s]\cdot[x_s]$$

$$\frac{d}{dt}M_L=0\cdot M_L;\ M_L=1\cdot M_L$$

(2.84)

where $[x_s]$ is the state vector and $[z]$ is the output vector of the disturbance system. A weighting matrix involving the difference between the measured output and the observer output is added to the final design, which is called the gain matrix of the observer. This matrix corrects the output of the model and improves the performance of the observer.

Out of expressions (2.82) and (2.84) the observer structure can be derived. In [60] it was demonstrated that the stationary error in the observed rotor position $\hat{\gamma}$ and the observed velocity $\hat{\omega}$ presented can be eliminated by adding a PI in the feedback loop of the resulting observer as depicted in Fig. 2.14. This configuration has been used in this work. Defining the observer error as the difference between the estimated and the observed values of the system, the following expressions are obtained:

$$\tilde{\omega} = \omega - \hat{\omega}, \quad \tilde{\gamma} = \gamma - \hat{\gamma}, \quad \tilde{M}_L = M_L - \hat{M}_{L0} . \tag{2.85}$$

The equations of the observer with PI in the feedback loop are:

$$\frac{d}{dt}\hat{\omega} = \frac{p}{J}M_i - \frac{p}{J}\hat{M}_{L0} - r_1, \quad r_1 = K_{PI} \cdot (\gamma - \hat{\gamma}) + \frac{K_{PI}}{T_{I1}} \cdot \int (\gamma - \hat{\gamma}) \cdot dt$$

$$\frac{d}{dt}\hat{\gamma} = \hat{\omega} + r_2, \quad\quad\quad r_2 = K_{P2} \cdot (\gamma - \hat{\gamma}) \tag{2.86}$$

By merging (2.82), (2.85) and (2.86) the following differential equations for the observer errors $\tilde{\omega}$ and $\tilde{\gamma}$ results:

$$\frac{d}{dt}\tilde{\omega} = \frac{p}{J}\tilde{M}_L - K_{PI} \cdot \tilde{\gamma} - \frac{K_{PI}}{T_{I1}} \cdot \int \tilde{\gamma} \cdot dt$$

$$\frac{d}{dt}\tilde{\gamma} = \tilde{\omega} - K_{P2} \cdot \tilde{\gamma} \tag{2.87}$$

Transforming the continuous-time expressions to the frequency domain, the transfer functions of $\frac{\tilde{\omega}(s)}{\tilde{M}_L(s)}$ and $\frac{\tilde{\gamma}(s)}{\tilde{M}_L(s)}$ can be obtained which are expressed as:

$$\frac{\tilde{\omega}(s)}{\tilde{M}_L(s)} = -\frac{p}{J} \cdot \frac{s^2 + K_{P2} \cdot s}{s^3 + K_{P2} \cdot s^2 + K_{PI} \cdot s + K_{I1}}$$

$$\frac{\tilde{\gamma}(s)}{\tilde{M}_L(s)} = -\frac{p}{J} \cdot \frac{s}{s^3 + K_{P2} \cdot s^2 + K_{PI} \cdot s + K_{I1}} \tag{2.88}$$

where $\tilde{\omega}(s)$, $\tilde{\gamma}(s)$ and $\tilde{M}_L(s)$ are the Laplace transforms of the considered output signals and of the input signal, respectively, and K_{I1} is defined as the quotient between the proportional controller gain K_{PI} and the integration time constant T_{I1}. Then, the parameters of the observer are determined by the poles of the transfer function (2.88). The three poles of the transfer function are equal to:

$$\beta_1 = \beta_2 = \beta_3 \Rightarrow s^3 + K_{P2} \cdot s^2 + K_{PI} \cdot s + K_{I1} = (s - \beta)^3 \Rightarrow \begin{matrix} K_{P2} = -3 \cdot \beta \\ K_{PI} = 3 \cdot \beta^2 \\ K_{I1} = -\beta^3 \end{matrix} \tag{2.89}$$

The selection of the best values of β is obtained after evaluating the resulting performances of the observer in both simulation and experimental tests. The value of β was chosen to fulfill a compromise between the desired dynamic of the observer and noise rejection of the velocity signal.

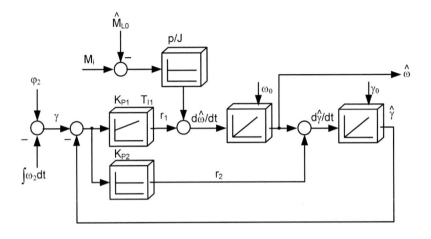

Fig. 2.14   Formal structure of the observer with PI feedback.

2.7   Summary of chapter 2

In this chapter, a mathematical model for the analytical description of squirrel cage IMs including their behavior in both healthy and faulty conditions is presented. The modelling is based on the coupled magnetic circuit theory and matrix notation is employed. Three possible representations of a broken bar are given. The one based on the principle of superposition although only valid for steady-state conditions, allows a comprehensive analysis of the effects of the rotor asymmetry in the response of the machine as will be shown in the next chapter. The other two representations, which are valid for dynamic analysis, differ in the necessary computational effort. In fact, for the case of one broken rotor bar, a reduction in a factor one in the order of the resulting system can be achieved. Nowadays, due to the available computing power, this is no longer a real advantage any more and thus, both representations are considered to be comparable.

Finally, the principle of field oriented control is presented. For the implementation of this control, the well known dynamic equations of the IM based on the space phasor theory and expressed in complex notation are formulated. Several variants of this control scheme which differ on the rotor flux model employed and on the utilization or not of angular transducers are analyzed. The approaches used in this work to determine the rotor flux angle φ_2 are based on the current/angle-model and the voltage/current-model. An encoderless variant of field oriented control is also described which makes use of the latter model for determining the rotor flux angle and an observer for obtaining the velocity signal. All schemes constitute state of the art for controlling IM. Its performance and behavior in the range of normal speed under the presence of rotor asymmetries is studied in the following chapter.

3 Effects caused by rotor asymmetries in electric drives

3.1 Effects of rotor asymmetries in case of ideal sinusoidal voltage supply

In the previous chapter a simplified approach of the model of IMs including rotor asymmetries based on the principle of superposition was explained. This model basically considers the faulty machine to be equivalent to the superposition of two configurations, a healthy and a faulty one (see Fig. 2.5). Then, an equivalent space phasor of rotor flux can be obtained out of the resulting flux distribution in the rotor cage. This space phasor, which will not describe a circular trajectory proper of balanced systems, can be obtained from the application of symmetrical components theory. According to this theory, an unbalanced system can be represented by one three-phase system rotating forward, one three-phase system rotating backwards and one system that does not rotate, which is also called zero sequence system [44, 46].

In healthy conditions, the rotor flux and current space phasors can be represented in steady-state by a system rotating at a frequency $s \cdot f_1$ forward with respect to a system rotating aligned to the rotor, being f_1 the stator voltage frequency and s the slip. Conversely, in case of one rotor asymmetry, the resulting space phasor can be represented by adding a phasor that rotates backwards at $-s \cdot f_1$. Then, in the stationary coordinate system depicted in Fig. 3.1, the resulting rotor space phasor $\underline{\psi}'_2$ is equal to

$$\underline{\psi}'_2 = \psi_{2,\text{forward}} \cdot e^{j(s \cdot \omega_1 + \dot{\gamma})t} + \psi_{2,\text{backward}} \cdot e^{j(-s \cdot \omega_1 + \dot{\gamma})t} \tag{3.1}$$

assuming that the rotor of the machine rotates at angular velocity $\dot{\gamma}$. Note that in steady-state, the following relationship between the slip angular frequency ω_2 defined in (2.73) and the angular frequency of the stator voltage $\omega_1 = 2 \cdot \pi \cdot f_1$ yields

$$\omega_2 = s \cdot \omega_1. \tag{3.2}$$

Then, the forward component gives rise to the fundamental component of the rotor flux, whereas the backward component generates a space phasor that rotates forward in this coordinate system but with a different velocity. Next, by substituting the rotational speed of the fundamental component of the flux space phasor $\dot{\phi}_2$ in the expression (3.1), it yields

$$\underline{\psi}'_2 = \psi_{2,\text{forward}} \cdot e^{j\Phi_2 t} + \psi_{2,\text{backward}} \cdot e^{j(\Phi_2 - 2 \cdot s \cdot \omega_1)t} \quad . \tag{3.3}$$

Thus, the component obtained with the MCSA becomes evident. This harmonic component also appears in the other variables of the machine such as the rotor current. In steady-state, the rotational frequencies of the rotor and stator flux space phasors coincide and thus the same considerations can be extended to the stator current and flux space phasors. Taking now into account that field oriented control uses the rotor flux angle φ_2 for the orientation, depicted in Fig. 3.1, it becomes clear that the fault signature will be present at some extent in the variables of the controllers, too. In fact, the fundamental component will vanish to be replaced by a DC component, and the harmonic signature at twice the slip frequency will be preserved.

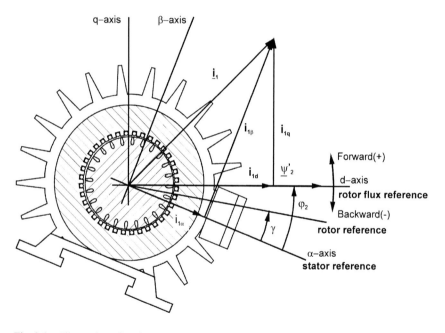

Fig. 3.1   Illustration of stationary coordinate system (α,β) and field coordinate system (d,q).

Because the fault signature depends on the operating point, the forthcoming analysis will be carried out for nominal operating conditions. Considering the usual values of nominal slip frequencies of IMs, the harmonic signature will be in the range of some Hertz.

3.2 Effects of rotor asymmetries in inverter operation with field oriented control

From the previous section it has been determined that the rotor asymmetry causes characteristic harmonics in the main quantities of the machine as currents and fluxes of the rotor as well as of the stator. In this section, it will be explained how this asymmetry affects the overall behavior of the drive in the case of field oriented control.

3.2.1 Fault propagation in the case of field oriented control with superimposed velocity control

For the next analysis it will be assumed that the machine is symmetric and the impact of one single broken rotor bar will be considered to be a perturbation. Besides it will be supposed that the flux calculated by the model has the same orientation that the rotor flux space phasor of the machine. Though the last assertion is not true in a strict sense, simulations have shown that the difference is in normal operating conditions very small and therefore negligible. Under these assumptions the stator current space phasor results from adding two terms: one term equals the value at healthy conditions and the second term which corresponds to the harmonic component caused by the fault denoted in this case with the subscript z. Then, the components of the stator current space phasor $\underline{i}_1$ in field coordinates are equal to

$$\left. \begin{array}{l} i_{1d}^{1bb} = i_{1d} + i_{1dz} \\ i_{1q}^{1bb} = i_{1q} + i_{1qz} \end{array} \right\} \tag{3.4}$$

In a similar way, this representation can be extended to the rotor flux space phasor $\underline{\psi}'_2$, the stator flux space phasor $\underline{\psi}_1$ and the rotor current space phasor $\underline{i}'_2$. On the other hand, in the the field coordinate system (d,q), rearranging the stator voltage equation (2.64) as a function of the stator current space phasor $\underline{i}_1$ and the rotor flux space phasor $\underline{\psi}'_2$ yields:

$$\underline{u}_1 = \left(R_1 + \frac{R'_2}{(1+\sigma_2)^2} \right) \cdot \underline{i}_1 + \sigma \cdot L_1 \cdot \frac{d\underline{i}_1}{dt} + j \cdot \sigma \cdot \dot{\varphi}_2 \cdot L_1 \cdot \underline{i}_1 - \frac{R'_2 \cdot \underline{\psi}'_2}{(1+\sigma_2)^2 \cdot L_{1h}} + \frac{j \cdot \dot{\gamma} \cdot \underline{\psi}'_2}{1+\sigma_2} \tag{3.5}$$

and splitting into real and imaginary parts, the stator voltage components result equal to

$$u_{1d} = R \cdot \left(i_{1d} + T_{1\sigma} \cdot \frac{di_{1d}}{dt} \right) + e_{1d}$$

$$u_{1q} = R \cdot \left(i_{1q} + T_{1\sigma} \cdot \frac{di_{1q}}{dt} \right) + e_{1q}$$

$$(3.6)$$

where

$$T_{1\sigma} = \frac{\sigma \cdot L_1}{R}, \quad R = R_1 + \frac{R'_2}{\left(1+\sigma_2\right)^2} \tag{3.7}$$

being $T_{1\sigma}$ the time constant and the inverse of R the gain $K_{1\sigma}$ of the first order complex system that represents the electrical behavior of the motor as shown in Fig. 3.2; and

$$e_{1d} = -\sigma \cdot \dot{\phi}_2 \cdot L_1 \cdot i_{1q} - \frac{R'_2}{\left(1+\sigma_2\right)^2} \cdot \frac{\psi_{2d}}{L_{1h}}$$

$$e_{1q} = \sigma \cdot \dot{\phi}_2 \cdot L_1 \cdot i_{1d} + \dot{\gamma} \cdot \frac{1}{1+\sigma_2} \cdot \psi_{2d}$$

$$(3.8)$$

depicted in the same figure. It is to be remarked that as stated in (2.70), the imaginary part of the rotor flux in field coordinates is zero by definition. And the augmented simplified machine model in field coordinates for a machine with an asymmetry is given by

$$u_{1d} = R \cdot \left(i_{1d} + T_{1\sigma} \cdot \frac{di_{1d}}{dt} \right) + e_{1d} + e_{1dz}$$

$$u_{1q} = R \cdot \left(i_{1q} + T_{1\sigma} \cdot \frac{di_{1q}}{dt} \right) + e_{1q} + e_{1qz}$$

$$(3.9)$$

being e_{1d} and e_{1q} the back-emfs, and e_{1dz} and e_{1qz} the terms generated by the faulty rotor which are derived from the simplified model of the faulty IM and are equal to

$$e_{1dz} = R \cdot \left(i_{1dz} + T_{1\sigma} \cdot \frac{di_{1dz}}{dt} \right) - \sigma \cdot \dot{\phi}_2 \cdot L_1 \cdot i_{1qz} - \frac{R'_2}{\left(1+\sigma_2\right)^2} \cdot \frac{\psi'_{2z}}{L_{1h}}$$

$$e_{1qz} = R \cdot \left(i_{1qz} + T_{1\sigma} \cdot \frac{di_{1qz}}{dt} \right) + \sigma \cdot \dot{\phi}_2 \cdot L_1 \cdot i_{1dz} + \dot{\gamma} \cdot \frac{1}{1+\sigma_2} \cdot \psi'_{2z}$$

$$(3.10)$$

Therefore (3.9) is obtained from applying the principle of superposition, and thus assuming linearity in the modeling of this phenomenon. The effects of the rotor asymmetry e_{1dz} and e_{1qz}

will be considered as disturbances in the control loop, which are depicted in Fig. 3.2. Because of digital sampling and the inverter, a delay T_0 is introduced in the current loop as shown in Fig. 3.2. This delay is normally replaced by a first order system with an equivalent time constant equal to $(1.5 \cdot T'_0)$, where T'_0 is the inverse of the switching frequency in an inverter controlled by space phasor modulation. Then, the PI controller in the current loop can be tuned according to the amplitude optimum method, provided that $T_0 \ll T_{1\sigma}$ [62], [63]. By using this design procedure, the integration time constant T_{Ii} and the proportional controller gain K_{Pi} are given by

$$T_{Ii} = T_{1\sigma}, \qquad K_{Pi} = \frac{T_{1\sigma}}{\left(2 \cdot T_0\right)}. \qquad (3.11)$$

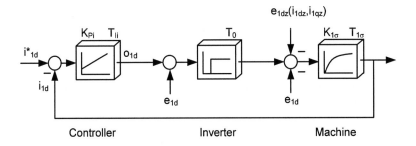

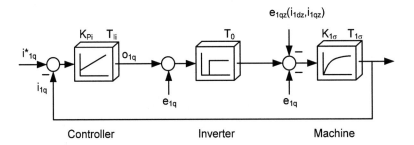

Fig. 3.2   Block diagram of the current control loops for an IM with a single broken bar.

On the other hand, several applications for field oriented control use a superimposed velocity control which is responsible for the generation of the torque producing current reference i^*_{1q}. Since the rotor asymmetry clearly influences the electromagnetic flux of the machine and therefore the torque, it can be expected that the velocity will also contain the characteristic oscillation to some extent. In this case, the effect of the rotor asymmetry can be represented by means of a disturbance M_z (i_{1dz}, i_{1qz}) in the electromagnetic torque, as it is shown in Fig 3.3 where the simplified block diagram of the velocity control loop is depicted, for an IM. Here, the current loop is approximated by a first order system [63]. Among the existing methods described in the literature to tune the velocity controller and obtain its parameters T_{In} and K_{Pn}, the Zieglers and Nichols method has been used in this work [62], [64].

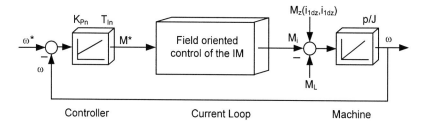

Fig. 3.3 Block diagram of the velocity control loop for an IM with a single broken bar.

In most drives the stator currents are available, so that they can be stored for further analysis. For this reason, currents will be one of the variables considered for the evaluation of the fault. On the other hand, the output of the controllers can provide valuable information for the diagnostics and its utilization is justified in the case that they can be easily accessed. Finally, the rotor flux will also be considered for the evaluation of the state of the IM.

3.2.2 Analysis of the suitable variables for the fault detection

Based on the previous simplified model of the inverter-fed asymmetric IM operating with field oriented control, an analysis of sensitivity with Bode diagrams is presented next in order to determine the effects of the rotor asymmetry on the main control variables.

3.2.2.1 Stator currents

If the reference torque producing current i^*_{1q} is considered to be constant, the effect of the disturbance in the torque producing current is given by the transfer function of i_{1q}/e_{1qz} which is equal to

$$G_{iq}(s) = \frac{i_{1q}(s)}{e_{1qz}(s)} = \frac{s \cdot K_{1\sigma} \cdot T_{li}}{s^3 \cdot T_{li}^2 \cdot T_0 + s^2 \cdot \left(T_0 + T_{li}\right) \cdot T_{li} + s \cdot \left(1 + K_{Pi} \cdot K_{1\sigma}\right) \cdot T_{li} + K_{Pi} \cdot K_{1\sigma}} \tag{3.12}$$

where $i_{1q}(s)$ and $e_{1qz}(s)$ are the Laplace transform of the output and input signals, respectively. The corresponding Bode diagram of (3.12) is depicted in Fig. 3.4, for an IM with the parameters given in the appendix 1 and where frequencies have been normalized with the stator fundamental frequency. For a properly tuned current loop, an attenuation of minimum -40dB can be obtained for disturbances in the current in a system which considers only this control loop. Furthermore, the capacity to reject these disturbances is even greater in the range of characteristic frequencies of the fault at nominal operating conditions. A typical slip value for low voltage IMs places the disturbance due to the fault at approx. 6 Hz, which according to Fig. 3.4 presents an attenuation in a factor equal to 813 (-58.2dB). The same analysis yields for the flux producing current provided that i^*_{1d} is constant.

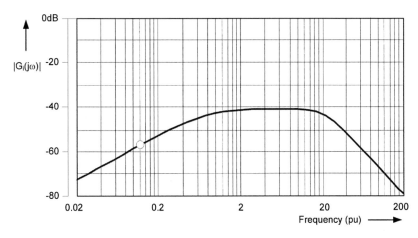

Fig. 3.4   Frequency response between the disturbance e_{1qz} and the current i_d or i_q. Frequencies are normalized with the fundamental frequency of the stator as base frequency.

On the other hand, the torque producing current i_{1q} usually depends on a superimposed velocity control as illustrated in Fig. 3.3. In this case, the torque disturbance M_z associated to the fault is visible in the output of the velocity controller M^*, that means in the i^*_{1q} quantity, with an amplitude which depends on the parameters of the velocity controller K_{Pn} and T_{In} and the inertia J and which is given by the following transfer function:

$$G_M(s) = \frac{M^*(s)}{M_z(s)} = \frac{(K_{Pn}/J)\cdot(1+s\cdot T_{In})\cdot(1+s\cdot T_1)}{(K_{Pn}/J)\cdot(1+s\cdot T_{In})+s^2\cdot T_{In}\cdot(1+s\cdot T_1)}. \tag{3.13}$$

By considering again an IM with the parameters described in the appendix 1, the resulting Bode diagram is depicted in Fig 3.5. It shows that the disturbance appears with almost unity gain in the output of the velocity controller, providing in this way a propagation path to the stator currents. However, for different inertia loads, the gain of the transfer function given by (3.13) at the range of frequencies of interest is reduced as the inertia J_N increases.

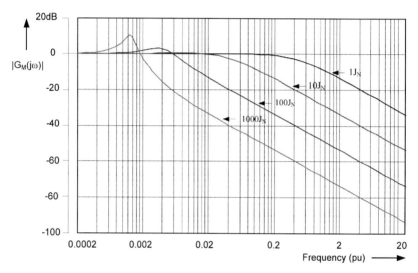

Fig. 3.5   Frequency response between the disturbance M_z and the output of the velocity controller M^* for different values of the inertia J_N. Frequencies are normalized with the fundamental frequency of the stator as base frequency.

3.2.2.2 Output of the current controllers

PI controllers reject the disturbance basically by generating a signal in the output with a similar frequency but a different phase. Assuming that the i^*_{1q} is constant, the transfer function between the disturbance e_{1qz} and the output of the controller o_{1q} (see Fig. 3.2) is given by:

$$G_{Rq}(s) = \frac{o_{1q}(s)}{e_{1qz}(s)} = \frac{K_{Pi}\cdot K_{1\sigma}\cdot(1+T_{1i}\cdot s)}{s^4\cdot T_{1i}^2\cdot T_0 + s^3\cdot(T_0+T_{1i})T_{1i} + s^2\cdot(1+K_{Pi}\cdot K_{1\sigma})T_{1i} + s\cdot K_{Pi}\cdot K_{1\sigma}} \tag{3.14}$$

whose Bode diagram considering again an IM with the parameters described in the appendix 1 is shown in Fig. 3.6. Then, the fault produces a disturbance in the current and this disturbance appears with unity gain in the output of the current controller o_{1q}, since the controller tries to compensate it. Therefore, the output of the controller o_{1q} is a suitable variable for fault detection provided that it is available. A similar analysis yields for the output of the flux producing current o_{1d}, also assuming constant reference i^*_{1d}. Furthermore, using o_{1d} as variable for fault detection is advantageous because it is independent of the operating point of the machine, namely of the load conditions.

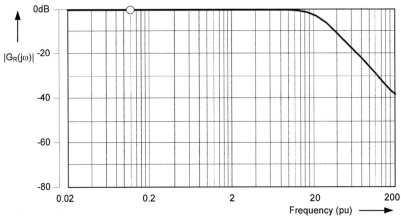

Fig. 3.6   Frequency response between the disturbance e_{1qz} and the output of the PI controller of i_d or i_q. Frequencies are normalized with the fundamental frequency of the stator as base frequency.

3.2.2.3   Rotor flux

In the following, the magnitude of the rotor flux space phasor based on the models described in subchapter 2.6.1 is evaluated in order to determine the suitability for the fault detection.

In the standard field oriented control, the current/angle-model is used and the magnitude of the rotor flux space phasor $\underline{\psi}'_2$ is obtained from (2.71) by passing the magnetizing current i_{1d} through a low pass filter with a time constant equal to T_2. Then, analogous to (3.12) the transfer function of $G_{id}(s) = i_{1d}(s)/e_{1dz}(s)$ can be obtained and the effect of the disturbance in the rotor flux is expressed as follows:

$$G_{\psi_2}(s) = \frac{\psi'_2(s)}{e_{1dz}(s)} = \frac{s \cdot K_{1\sigma} \cdot T_{li}}{s^3 \cdot T_{li}^2 \cdot T_0 + s^2 \cdot (T_0 + T_{li}) T_{li} + s \cdot (1 + K_{Pi} \cdot K_{1\sigma}) T_i + K_{Pi} \cdot K_{1\sigma}} \cdot \frac{1}{s \cdot T_2 + 1} \quad (3.15)$$

The corresponding Bode diagram for this transfer function is depicted in Fig. 3.7. It can be seen that the disturbance due to the fault appears in magnitude of the rotor flux space phasor highly attenuated.

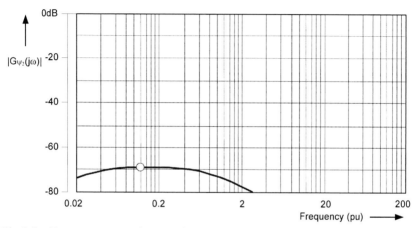

Fig. 3.7   Frequency response between the disturbance e_{1dz} and calculated rotor flux based on the current/angle-model. Frequencies are normalized with the fundamental frequency of the stator as base frequency.

Alternatively, a common approach to obtain the rotor flux angle is using the so called voltage/current-model that calculates the rotor flux using the voltages and currents. This case corresponds to a sensorless scheme since the rotor flux position is determined without measuring the rotor position (see subchapter 2.6.3). As it was previously discussed, the fault signature does not appear in the stator currents of the machine in the basic form of field oriented control without superimposed flux or velocity control theoretically if the references i^*_{1d} and i^*_{1q} are constant. However, the fault clearly propagates through the controllers to their respective outputs, which are determined by the voltage applied to the machine. Thus, it can be expected that the flux calculated in this way shows also the signature of the fault as it is next justified. The transfer function can be obtained out of (2.76) and (2.78) using the Laplace transformation:

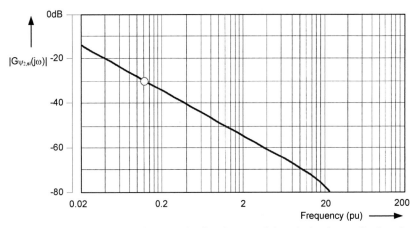

Fig. 3.8 Frequency response between the disturbance and the calculated rotor flux based on the voltage/current-model. Frequencies are normalized with the fundamental frequency of the stator as base frequency.

$$\psi'_{2,sl}(s) = \frac{\sigma \cdot L_{1h}}{\sigma - 1} \cdot i_{1d}(s) + (1 + \sigma_2)(u_{1d}(s) - R_1 \cdot i_{1d}(s)) \cdot \frac{1}{s} \qquad (3.16)$$

where the subscript "sl" is used to denote sensorless. Considering the transfer function $G_{id}(s) = i_{1d}(s)/e_{1dz}(s)$, then:

$$\psi'_{2,sl}(s) = \frac{\sigma \cdot L_{1h}}{\sigma - 1} \cdot G_{id}(s) \cdot e_{1dz}(s) + (1 + \sigma_2)(u_{1d}(s) - R_1 \cdot G_{id}(s) \cdot e_{1dz}(s)) \cdot \frac{1}{s}. \qquad (3.17)$$

And if the back-emf e_{1d} and e_{1q} of the control loop and the delay time T_0 shown in Fig. 3.2 are neglected, the transfer function $G_{Rd}(s)$ given by (3.14) yields for the voltage u_{1d} and then it is obtained that

$$\psi'_{2,sl}(s) = \frac{\sigma \cdot L_{1h}}{\sigma - 1} \cdot G_{id}(s) \cdot e_{1dz}(s) + (1 + \sigma_2)(G_{Rd}(s) \cdot e_{1dz}(s) - R_1 \cdot G_{id}(s) \cdot e_{1dz}(s)) \cdot \frac{1}{s}. \qquad (3.18)$$

Finally, rearranging terms, the following transfer function between the disturbance e_{1dz} and the rotor flux $\psi'_{2,sl}$ is obtained

$$G_{\psi_{2,sl}}(s) = \frac{\psi'_{2,sl}}{e_{1dz}(s)} = \frac{\sigma \cdot L_{1h}}{\sigma - 1} \cdot G_{id}(s) + \frac{(1 + \sigma_2)}{s} \cdot (G_{Rd}(s) - R_1 \cdot G_{id}(s)). \qquad (3.19)$$

Next, assuming that the references i^*_{1d} and i^*_{1q} are constant, the influence of $G_{id}(s)$ in the frequency of interest can be neglected as it was illustrated in Fig. 3.4. Therefore the transfer function can be simplified as follows:

$$G_{\psi_{2,sl}}(s) = (1+\sigma_2) \cdot \frac{G_R(s)}{s}. \tag{3.20}$$

The resulting Bode diagram of this transfer function is presented in Fig. 3.8. It is shown that the fault signature appears in the flux with a higher magnitude than in the previous case given by (3.15). With the optimal tuning of the current controllers, a minimum attenuation of -15dB is obtained which increases with the frequency.

3.3   Summary of chapter 3

In this chapter, the effects caused by rotor asymmetries in the behavior of an IM drive have been described and theoretically analyzed for steady-state operation.

A broken bar represents a local asymmetry in the cage rotor construction and thus a local perturbation of the air gap MMF. Based on the superposition of the healthy and faulty configurations of the IM, the appearance of a characteristic harmonic in the main quantities of the machine for the ideal case of sinusoidal voltage supply is justified.

To extend this analysis to the case of inverter operation with field oriented control, a simplified model is introduced where the machine is assumed to be symmetrical and the impact of the fault is added as a perturbation. Then, the effects of the asymmetry on the different control variables are determined with the analysis of sensitivity in the corresponding Bode diagrams.

For closed-loop inverter drives, in the ideal case of having a constant reference i^*_{1q}, disturbances in the current in the range of a few Hertz are attenuated provided that the current controllers are properly tuned. The same yields for the flux producing current. On the other hand, if there is a superimposed velocity control, the output of the velocity controller constitutes a propagation path to the current, although in an amount which depends on the total inertia. In this case, the torque producing current i_{1q} can give enough information for the diagnostics if the characteristic harmonic introduced by the asymmetry also appears in the velocity, as it occurs in the case of medium or low inertia drives. Alternatively, the outputs of the controllers o_{1d} and o_{1q} can be used for the diagnostics in this case, since in their corresponding Bode diagram the disturbance appears with unity gain in the output of the current controllers.

Finally, the magnitude of the rotor flux space phasor $\underline{\psi}'_2$ based on the current/angle-model and on the voltage/current-model is analyzed. Only in the case of using the latter model, the magnitude of the rotor flux space phasor represents a suitable variable for the fault detection. Conversely to the torque producing current, this variable exhibits the characteristic harmonic at low load.

4 Detection of rotor asymmetries

4.1 Introduction

This chapter introduces a new method for the detection of rotor asymmetries in field oriented controlled IMs based on the Fourier series. As previously demonstrated, rotor asymmetries introduce characteristic frequencies in the terminal variables of the IM which propagate to the rest of control variables. The standard method for the implementation of Fourier analysis in Digital Signal Processors (DSP) is the Discrete Fourier Transform (DFT). Since for the present application the determination of the whole spectrum is not needed, alternative methods to the fast Fourier Transform (FFT) algorithm are analyzed. These are the Goertzel's and the Sliding window DFT (SDFT) algorithms, whose principles are next presented. Based on the SDFT, a new fault indicator is proposed and main issues for its practical implementation are described in detail. As it will be explained, it uses an angle as integration variable instead of the time and it performs properly even at quasi-stationary conditions with variable frequency.

4.1.1 Discrete-time Fourier Transform

A common approach to detect a harmonic component uses the discrete-time Fourier transform (DTFT). The DTFT of a sequence x(n) is defined [66] by

$$X(e^{j\omega}) = \sum_{n=-\infty}^{\infty} x(n) \cdot e^{-j\omega n} \tag{4.1}$$

where ω is the real frequency variable. In the case of a finite-length sequence x(n) defined for $0 \leq n \leq N\text{-}1$, there is a simpler relation between the sequence and its DTFT $X(e^{j\omega})$ which consists in uniformly sampling the continuous function $X(e^{j\omega})$ between $0 \leq \omega \leq 2\pi$ on the ω-axis at values $\omega = \dfrac{2\pi k}{N}$ with $0 \leq k \leq N\text{-}1$. From (4.1),

$$X(k) = X(e^{j\omega})\Big|_{\omega=\frac{2\pi k}{N}} = \sum_{n=0}^{N-1} x(n) \cdot e^{-j\frac{2\pi k}{N} n} \tag{4.2}$$

where k is a frequency index and the factor $e^{-j\frac{2\pi k}{N}n}$ represents a complex function rotating clockwise in the complex plane with an angular velocity of ω. The resulting sequence X(k) is called the DFT of x(n). Conversely to the DTFT which only exists for an indefinite continuous range of digital frequencies, the DFT is generated by a finite sum and it is defined for a finite set of frequency values. With the following notation

$$W_N = e^{-j\frac{2\pi}{N}}, \tag{4.3}$$

equation (4.2) can be expressed in a more compact form as

$$X(k) = \sum_{n=0}^{N-1} x(n) \cdot W_N^{kn}, \quad 0 \le k \le N-1. \tag{4.4}$$

The DFT provides a practical approach to the numerical computation of the DTFT of a finite-length sequence, particularly if efficient algorithms are used of the computation of the DFT as explained below.

4.1.2 Goertzel's algorithm

The Goertzel's algorithm constitutes a common digital processing technique for those applications where only a few DFT frequencies of the total spectrum are needed [67]. It is usually implemented in the form of a second order infinite impulse response (IIR) filter that computes a single term of the DFT output, i.e. the harmonic component k^{th} of an N-point DFT defined by (4.2). The Goertzel's algorithm exploits the periodicity of the phase of the Euler function [65] which gives the following identity valid for all integer values of k

$$W_N^{-kN} = 1. \tag{4.5}$$

Using this identity, the expression (4.2) can be rewritten as

$$X(k) = \sum_{m=0}^{N-1} x(m) \cdot W_N^{km} = W_N^{-kN} \cdot \sum_{m=0}^{N-1} x(m) \cdot W_N^{km} = \sum_{m=0}^{N-1} x(m) \cdot W_N^{-k(N-m)}. \tag{4.6}$$

For the sake of clarity, the variable y(n) will be introduced which delivers the output at $X(k) = y_k(n)|_{n=N}$. Then (4.6) is expressed as

$$y_k(n) = \sum_{m=0}^{n} x(m) \cdot W_N^{-k(n-m)} \tag{4.7}$$

and by replacing m'= n-m, it yields

$$y_k(n) = \sum_{m'=0}^{n} x(n-m') \cdot W_N^{-km'} \tag{4.8}$$

and this equation can be written in a recursive form as follows

$$y_k(n) = x(n) + x(n-1) \cdot W_N^{-k} + x(n-2) \cdot W_N^{-k2} + x(n-3) \cdot W_N^{-k3} + \cdots + x(0) \cdot W_N^{-kn}, \tag{4.9}$$

or equivalently in a more compact form

$$y_k(n) = x(n) + y_k(n-1) \cdot W_N^{-k}. \tag{4.10}$$

From expression (4.10), the corresponding filter structure can be derived from

$$\begin{aligned}
y_k(n) &= \frac{x(n)}{\left(1 - z^{-1} \cdot W_N^{-k}\right)} = \frac{x(n) \cdot \left(1 - z^{-1} \cdot W_N^{k}\right)}{\left(1 - z^{-1} \cdot W_N^{-k}\right) \cdot \left(1 - z^{-1} \cdot W_N^{k}\right)} \\
&= \frac{\left(1 - z^{-1} \cdot W_N^{k}\right) \cdot x(n)}{\left(1 - 2 \cdot z^{-1} \cdot \cos\left(\frac{2\pi k}{N}\right) + z^{-2}\right)}
\end{aligned} \tag{4.11}$$

Thus, the z-domain transfer function of the Goertzel's filter is equal to:

$$G(z) = \frac{\left(1 - z^{-1} \cdot W_N^{k}\right)}{\left(1 - 2 \cdot z^{-1} \cdot \cos\left(\frac{2\pi k}{N}\right) + z^{-2}\right)}. \tag{4.12}$$

This transfer function can be divided in two parts, which are

$$v_k(n) = \frac{x(n)}{\left(1 - 2 \cdot z^{-1} \cdot \cos\left(\frac{2\pi k}{N}\right) + z^{-2}\right)} \tag{4.13}$$

and

$$y_k(n) = \left(1 - z^{-1} \cdot W_N^{k}\right) \cdot v_k(n) \tag{4.14}$$

which give rise to the two recursive equations of Goertzel. It must be remarked that the output of the filter $y_k(n)$ delivers the DFT output coefficient $X(k)$ only at the time index n=N, where k is an integer pertaining to the range $0 \le k \le N-1$. For the practical implementation, equations (4.13) and (4.14) can be expressed in a more useful way by means of their corresponding difference equations as

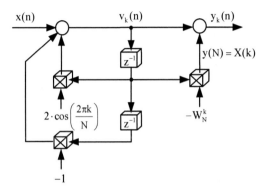

Fig. 4.1   Block diagram of the IIR filter for the implementation of Goertzel's algorithm

$$v_k(n) = 2 \cdot \cos\left(\frac{2\pi k}{N}\right) \cdot v_k(n-1) - v_k(n-2) + x(n),$$

$$y_k(n) = v_k(n) - W_N^k \cdot v_k(n-1)$$

(4.15)

which are graphically represented in Fig. 4.1. The advantage of implementing the Goertzel's filter compared to an N-point $X(k)$ DFT for a single harmonic component, is that $v_k(n)$ must be calculated N times, while $y_k(n)$ needs only to be computed once after the arrival of the Nth input sample. Then, processing of a signal $x(n)$ requires just N+2 real products and 2N +1 real additions in order to obtain the DFT output coefficient $X(k)$, i.e. a single harmonic component at the time index n=N. As a result, this algorithm provides a significant computational advantage to the direct use of the definition equation of the DFT. However, similar to the DFT, it cannot compute Fourier coefficients until processing all the samples corresponding to a complete period. The number of samples equals the transformation length and the detection time increases as this length does. Because of this limitation this algorithm is not implemented in this work. Instead a new algorithm which is described in the next subchapter will be used.

4.1.3 Sliding window DFT Algorithm

The Sliding DFT (SDFT) algorithm is especially efficient for narrowband spectrum analysis, as it occurs with the Goertzel algorithm [68]. As the name indicates, for a given harmonic component the algorithm calculates its corresponding DFT by using a moving window. Since the Fourier coefficients are updated at every sampling time, this algorithm delivers a continuous output value of the computed variable and is thus different to the Goertzel's algorithm, in which the result is obtained after a period of computation. Thus it is suitable for the analysis of non-stationary cases. In addition it results in a more efficient algorithm in terms of computational effort, as demonstrated next.

In order to clarify the operation principle, the waveform of Fig. 4.2 will be used. Initially, the algorithm calculates the DFT using N samples. Then, the window is shifted in one sample, considering a new N-point sample set. This fact permits the use of the DFT shifting theorem [69] to simplify the calculations of the new DFT harmonic component. The shifting theorem states that the kth harmonic component of the DFT of a finite length windowed sequence $X_k(n)$ shifted by one sample is equivalent to $X_k(n) \cdot W_N^{-k}$.

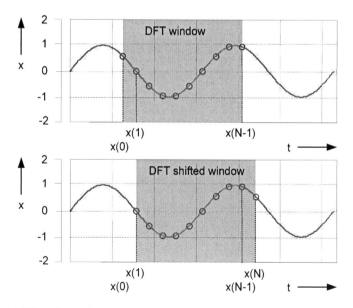

Fig. 4.2   Sliding window for DFT calculation.

4.1.3.1 Main algorithm

From (4.2), the kth harmonic component of a N-sample DFT can be defined as

$$X_k = \sum_{m=0}^{N-1} x(m) \cdot W_N^{km} \tag{4.16}$$

Since in this case is also relevant the sampling instant when the DFT is calculated, (4.16) will be expressed as a function of the sampling instant n, being $n \geq N-1$

$$X_k(n) = \sum_{m=0}^{N-1} x(n+m-N+1) \cdot W_N^{km}. \tag{4.17}$$

Then, expanding last equation the following expression can be obtained

$$X_k(n) = x(n+1-N) + x(n+2-N) \cdot W_N^k + x(n+3-N) \cdot W_N^{k2} + \cdots \\ + x(n-1) \cdot W_N^{k(N-2)} + x(n) \cdot W_N^{k(N-1)}. \tag{4.18}$$

In a similar way, the kth harmonic component at the previous sampling time equals

$$X_k(n-1) = x(n-N) + x(n+1-N) \cdot W_N^k + x(n+2-N) \cdot W_N^{k2} + \cdots \\ + x(n-2) \cdot W_N^{k(N-2)} + x(n-1) \cdot W_N^{k(N-1)}. \tag{4.19}$$

and after multiplying (4.19) by W_N^{-k}, it results

$$X_k(n-1) \cdot W_N^{-k} = x(n-N) \cdot W_N^{-k} + x(n+1-N) + x(n+2-N) \cdot W_N^k + \cdots \\ + x(n-2) \cdot W_N^{k(N-3)} + x(n-1) \cdot W_N^{k(N-2)}. \tag{4.20}$$

Then, by subtracting (4.18) from (4.20) it yields

$$X_k(n) - X_k(n-1) \cdot W_N^{-k} = x(n) \cdot W_N^{(N-1)k} - x(n-N) \cdot W_N^{-k}. \tag{4.21}$$

Thus,

$$X_k(n) = W_N^{-k} \cdot \left(X_k(n-1) + x(n) - x(n-N) \right). \tag{4.22}$$

This equation gives a very straightforward method to calculate the DFT, where $X_k(n-1)$ is phase shifted by W_N^{-k}, then subtracted with the x(n-N) sample and added with new sample

72

x(n). The resulting amplitude $|X_k(n)|$ will be applied for the detection of rotor asymmetries based on the torque producing current, as it will be explained in the next subchapter.

The output $X_k(n)$ will only be valid after having processed N samples in the input, but just few computations for the next value $X_k(n+1)$ are needed. It is to be remarked that N can be any positive integer and thus the algorithm can be tuned at any frequency of interest. Then, the z-domain transfer function for the kth harmonic component of the SDFT is given by

$$G(z) = \frac{1 - z^{-N}}{W_N^k - z^{-1}} = \frac{z^N - 1}{z^{N-1}(z \cdot W_N^k - 1)}. \tag{4.23}$$

Since the filter response is truncated to N samples, the impulse response is finite in length, making it identical to the single frequency component DFT centered at normalize frequency of $\frac{2\pi k}{N}$. The filter structure for determining the DFT of a single harmonic component is depicted in Fig. 4.3.

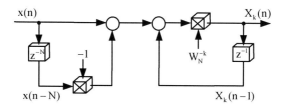

Fig. 4.3   Block diagram of the sliding window for DFT calculation.

4.1.3.2 Stability considerations

In practice, apart from the computational efficiency, the stability of the structure is very important due to recursive realization [71]. For the explained algorithm, the corresponding zero/pole diagram for k=2 and N=10 is presented in Fig. 4.4. This diagram shows that its poles are located in the unit circle. As an obvious consequence, the filter is marginally stable.

In the case that numerical rounding of the coefficients causes the filter to move out of the unit circle, a damping factor d can be added to force the pole to be inside of the stability region. The resulting transfer function is then described by

$$H(z) = \frac{z^N \cdot d^{-N} - 1}{z^{N-1}(z \cdot W_N^k \cdot d^{-1} - 1)}. \tag{4.24}$$

Then, by making d close to but less than 1 the error can be reduced. The selection of a suitable factor d depends on practical issues of the DFT implementation, such as the precision of the processor, being especially critical in fixed point implementations that are common in electrical drives.

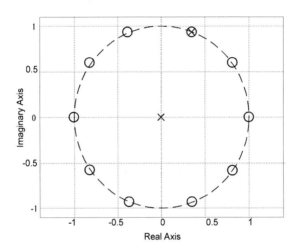

Fig. 4.4   Zero/pole locations for the sliding DFT.

4.1.3.3   Numerical performance

Table 2 shows the comparison in terms of performance and demand of computational resources of the three algorithms here presented, which constitute the main algorithms existing in the literature to obtain one DFT harmonic component using an N-length data array.

The first alternative, a normal DFT, requires 2N multiplications and 2N additions, as the real and imaginary parts must be considered. A new value of the DFT will be available only after N new samples, requiring the same calculation burden. On the other hand, the Goertzel's algorithm uses N real multiplications for the iterative equation (4.22) and 2 extra multiplications, one for the real and one for the imaginary part. The number of additions in the real part of (4.22) is 2N, and only one extra addition is needed in the imaginary part. The same number of calculations is needed for a new calculation of the harmonic component. Finally, the SDFT, uses 4N multiplications and 4N additions to determine the first value of the spectral component. After that, just 4 multiplications and 4 additions are needed for obtain a new value of the DFT harmonic component. As a result, the SDFT algorithm results clearly in the most advantageous approach in terms of computational efficiency and thus it is the selected algorithm to be used.

Table 2. Performance comparison of different algorithms.

	Single X_k computation		Next X_k computation	
	Multiplications	Additions	Multiplications	Additions
DFT	2N	2N	2N	2N
Goertzel	N+2	2N+1	N+2	2N+1
SDFT	4N	4N	4	4

4.2   Real time implementation of the proposed fault indicator

Different variables of the control loop are distorted by the rotor asymmetry and thus exhibit the characteristic harmonic component caused by that fault at twice the slip frequency, as explained in chapter 3. Though fault detection could be possible by simple visual inspection of the signal in the time domain, a transformation to the frequency domain facilitates the automatic detection. In principle, a FFT algorithm could be utilized for this task, but since the frequency of the searched component is a priori known, less computational demanding algorithms are preferred as explained in the subchapter 4.1.

In particular, the amplitude of the harmonic component $|X_k(n)|$ at twice the slip frequency will be obtained from expression (4.22) with k=1, being x(n) a variable distorted by the asymmetry as for example the torque producing current i_{1q}. First, the frequency of the characteristic harmonic is available since the slip frequency is an intermediate variable needed in some variants of field oriented control for determining the rotor flux angle as shown in expression (2.75) being equal to

$$\omega_2 = \frac{i_{1q}}{T_2 \cdot i_{\mu 2}}. \tag{4.25}$$

Then, the variable to obtain the finite-length sequence x(n) in (4.22) will be the angle θ generated directly from twice the slip frequency as

$$\theta(t) = 2 \cdot \int \omega_2 \cdot dt . \tag{4.26}$$

It is important to remark that even though the time does not explicitly appear, a window calculation over a period of the signal can be performed provided that the angle is known.

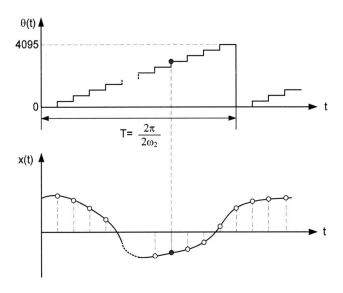

Fig. 4.5   Sampling of a variable x(t) exhibiting the component caused by the rotor asymmetry using the discrete changes of the angle θ.

In the control platform used in this work, the angles are defined as positive integers of 12 bits, where 0 corresponds to zero radians and 4095 to $\left(2\pi - \frac{4096}{2\pi}\right)$ rad, as shown in Fig. 4.5. Then, each time that the angle θ changes, the variable x(n) under evaluation is stored in an array of a length equivalent to the maximum value of θ, i.e. $N=2^{12}$. In other words, the angle θ will point to the most recent value x(n) and the angle θ+1 to the oldest stored value, x(n-N). Similarly, the amplitude of the harmonic component X(n) is stored in another array of the same size, so that the component X(n-N) is also obtained by pointing to θ+1. This permits a fast calculation of the term X(n). A simplified block diagram of the described detection algorithm is shown in Fig. 4.6. For the sake of simplicity, since in the following only the first harmonic component of (4.22) will be used, $X_{k=1}$ will be denoted as X.

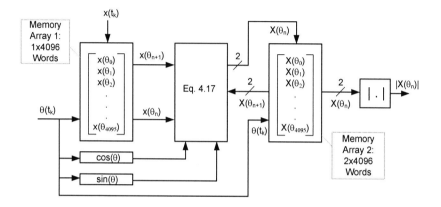

Fig. 4.6   Block diagram of the detection algorithm.

4.3   Performance of the proposed fault indicator

The new proposed indicator basically calculates the amplitude at the characteristic frequency caused by the fault in the input signal and the employed algorithm makes the indicator effective even during quasi-stationary conditions with variable slip frequency. Next, the performance of the indicator during transients and dynamic operation is discussed. As it will be explained, due to the nature of the investigated fault, it is feasible to use this indicator together with a procedure able to eliminate those false alarms that may be originated by the occurrence of an abrupt change in the operating conditions of the drive.

4.3.1   Behavior of the fault indicator during transients

Depending on the variable used for the fault detection, a change in the operating point of the machine may affect the fault indicator. The rotor flux is not strongly affected by a torque step, provided that the control is properly decoupled. On the other hand, if only the torque producing current i_{1q} is available for the diagnostics, fast torque steps can produce a transient as shown in Fig. 4.7.a) which may exceed the threshold of the fault detection. The threshold for the indicator can be obtained operating the IM in healthy conditions, so that a distinction between intrinsic asymmetries in the rotor due to manufacture and incipient faults as partially broken bars can be assured.

A very simple approach to avoid a false fault detection considers the use of a low pass filter to the signal used for the diagnostics, the i_{1q} in this case, in order to mitigate the effect of fast transients. Among the different existing types of low pass filter, low pass FIR filter are best suited for digital implementation [68]. In this way, independently of the state of the machine, the fault indicator will be smoothed each time a transient occurs as shown in Fig. 4.7.b). The selection of the time constant of this additional filter depends on the dynamics of the system, which is a priori known or can be approximately estimated for each application.

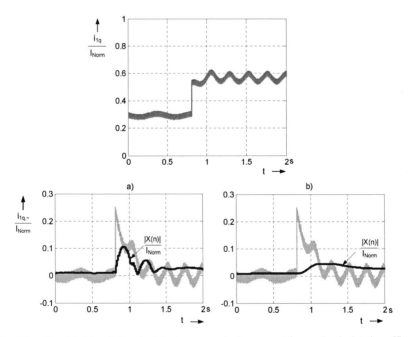

Fig. 4.7   Simulated i_{1q}, AC part of the torque producing current $i_{1q,\sim}$ and calculated coefficient $|X(n)|$ during a step load with the faulty IM: a) without filter, b) with filter.

4.3.2 Behavior of the fault indicator during dynamic operation

Dynamic operation obeys periodic patterns, also called duty cycles, of known period and magnitude. In general, the proposed fault algorithm would work even during dynamic operation. Nevertheless, a theoretical case exists which corresponds to a torque reference with identical frequency to that of the fault signature that would be critical. For all other cases, that is duty cycles with frequencies different from that originated by the rotor asymmetry the use of the proposed indicator is plausible.

4.3.3 Sensitivity

The characteristic harmonic due to the fault will be properly determined by the new proposed algorithm if its frequency matches the frequency calculated by the control algorithm, i.e. twice the slip frequency. In a practical implementation, this frequency could experiment small variations caused by transients and changes in the parameters of the IM model. Under these conditions the selection of the number of samples N is crucial. Besides the computing efficiency, the number of samples N influences the bandwidth of the indicator. A too narrow bandwidth can result in non proper detection of the characteristic frequency due to the fault, especially in the aforementioned cases. The optimal value of N can be determined with the help of the spectral diagram of the DFT algorithm.

If x(n) is the characteristic component of the fault consisting of a sampled sinusoid with arbitrary frequency ω_x

$$x(t) = \sin(\omega_x \cdot t) \tag{4.27}$$

and assuming that the DFT algorithm calculates the harmonic component at a frequency $\omega_k = \frac{2\pi k}{N \cdot T_s}$, where T_s is the sampling period and N the number of samples.

Substituting in (4.4), the following expression results

$$X(\omega_k) = \sum_{n=0}^{N-1} e^{j \cdot \omega_x \cdot n \cdot T_s} \cdot e^{-j \cdot \omega_k \cdot n \cdot T_s} = \sum_{n=0}^{N-1} e^{j(\omega_x - \omega_k) \cdot n \cdot T_s} = \frac{1 - e^{j(\omega_x - \omega_k) \cdot N \cdot T_s}}{1 - e^{j(\omega_x - \omega_k) \cdot T_s}}, \tag{4.28}$$

and the absolute value of this expression yields

$$|X(\omega_k)| = \left| e^{j\frac{(\omega_x-\omega_k)\cdot(N-1)\cdot T_s}{2}} \cdot \left(\frac{\sin\dfrac{(\omega_x-\omega_k)\cdot N\cdot T_s}{2}}{\sin\dfrac{(\omega_x-\omega_k)\cdot T_s}{2}} \right) \right| = \left| \frac{\sin\dfrac{(\omega_x-\omega_k)\cdot N\cdot T_s}{2}}{\sin\dfrac{(\omega_x-\omega_k)\cdot T_s}{2}} \right|. \qquad (4.29)$$

This equation indicates that the DFT behaves as a digital filter with a bandpass characteristic centered at frequency ω_k, as it is shown in Fig. 4.8.

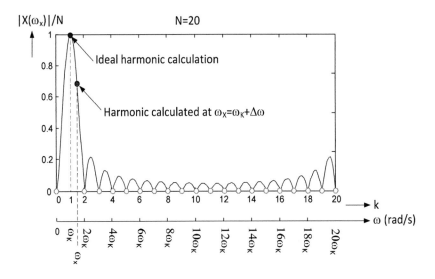

Fig. 4.8   Frequency response of a DFT with N=20. Graphical representation of equation (4.29) for an arbitrary frequency ω_x.

Then, the amplitude of the spectral component will be the maximum if ω_x exactly equals ω_k. This bandwidth should be made equal to several Hertz in order to cope with variations in the determination of ω_k in the order of one Hz.

Regarding the determination of the number of samples N for the proposed algorithm, contrary to other time domain based approaches, is always a fixed number per period of the signal. As a consequence, the DFT will deliver information about the harmonics at multiples of this frequency, i.e. ω_k, $2\omega_k$, $3\omega_k$,...$N\omega_k$ as it can be seen in Fig. 4.8. It must be remarked, that a too elevated value of N would need extra memory space for the processing of the SDFT. Therefore N should be kept as low as possible.

Since the SDFT basically samples the fault related variable at a lower frequency than the control frequency, an antialiasing filter should be added [70]. For the sake of simplicity, this filter will be implemented per software using a standard first order low pass filter with a cutoff frequency equal to the maximum frequency considered by the detection algorithm. If ω_{MAX}^{1bb} is used to denote this maximal frequency, then according to the Nyquist criterion it results to be equal to

$$\omega_{MAX}^{1bb} \le \frac{N}{2}\omega_k \; . \tag{4.30}$$

As this cutoff frequency of the antialising filter is fixed, the number of samples N of the detection algorithm should by high enough to comply with the Nyquist criterion also in the case that the fault presents the lowest frequency $\omega_k = \omega_{min}^{1bb}$, i.e. during operation under light load. Using this principle and (4.30), the following relationship yields

$$N \approx \frac{2 \cdot \omega_{MAX}^{1bb}}{\omega_{min}^{1bb}} . \tag{4.31}$$

Assuming for example that the lowest frequency to be detected is 10% of the maximum frequency, i.e. $\omega_{min}^{1bb} = 0.1 \cdot \omega_{MAX}^{1bb}$, then N is equal to 20. Furthermore, the antialiasing filter fulfills also the function of smoothing transients (see 4.3.1).

4.4   Summary of chapter 4

A new online diagnostics method for the detection of rotor asymmetries in field oriented controlled IMs based on the discrete Fourier transformation has been proposed. Since for this application the determination of the whole spectrum is not needed, alternative methods to the FFT algorithm are described like the Goertzel's and the SDFT algorithms and compared in terms of numerical performance.

The algorithm used here is based on the recursive DFT with a moving window and it requires a reduced number of samples and few calculations for the determination of the amplitude of the searched harmonic. One main feature of the proposal is the use of an angle as integration variable instead of the time. This angle is obtained from the integral of twice the slip frequency which is a characteristic variable available in the control scheme. Obviously a certain level of load is required. Then, according to subchapter 3.2, several signals exhibit the characteristic harmonic caused by the fault and thus the proposed method yields for all of them.

After explaining the issues dealing with the real time implementation of the proposed algorithm, its performance during transients and dynamic operation was discussed. Indeed, unlike other diagnostics methods, it works properly even at quasi-stationary conditions when the slip frequency varies. Regarding the former, a simple approach has been proposed consisting in the addition of a suitable low pass filter. This filter eliminates those false alarms in the detection originated by abrupt changes in the operating conditions of the drive. Furthermore a threshold for the indicator based on measurements of the IM in healthy conditions would be needed in order to discriminate between intrinsic asymmetries in the rotor due to manufacture and incipient faults as partially broken bars. During dynamic operation, only the theoretical case in which the torque pulsates exactly at the frequency of the characteristic harmonic due to the fault, could detection based on this approach be critical.

Finally, the influence that small variations in the slip frequency may cause in the performance of the detection algorithm was analyzed. It was demonstrated that a suitable number of samples N have to be used to attain the necessary bandwidth of the indicator for the fault detection.

5   Experimental results

5.1   Experimental set-up

Two identical IMs are used in the test bench, as can be seen in Fig. 5.1.a). One machine is used as a reference and the other one is prepared with an asymmetric rotor. In essence, drilling a hole into a rotor bar with its diameter and height allows the artificial creation of a single broken bar in an aluminum die-cast rotor. For the IM of study, in order to ensure that the current flowing through the bar is totally interrupted; two holes are drilled into the bar instead. The rotor with the built-in asymmetry is shown in Fig. 5.1.b).

a)   b)

Fig. 5.1   Test bench: a) tested IMs and b) detail of the asymmetric rotor.

The torque level is set with the two configurations shown in Fig. 5.2. Letter "A" denotes the configuration that uses an IM operated with an industrial drive with a braking chopper module. Letter "B", also in Fig. 5.2, denotes the second configuration consisting of a permanent magnet synchronous machine with a variable resistor. Experimental results are obtained with two different platforms for the control. One platform is based on the Vecon system [60] and the second one is based on a prototyping platform dSpace 1104. This second platform facilitates the monitoring of the different variables of the drive and the implementation of the fault detection algorithms developed in the previous chapter. In addition, it allows tests with the two additional variants of field oriented control investigated in this work, namely the field oriented control that uses the voltage/current-model of the rotor flux (Fig. 2.10) and the encoderless variant (Fig. 2.13). A 2048 pulses incremental encoder attached to the shaft of the machine is used to measure the rotor position, for those variants of the field oriented control requiring this variable. A simplified scheme of the complete set-up is depicted in Fig. 5.2. Regarding the prototyping platform dSpace 1104, the main characteristics are:

- High level programming using Matlab Simulink and a C compiler
- The platform includes a Power PC 603e processor with a slave TMS320C200 DSP controller for I/O task
- 8 A/D converters, four of them are 16 bits converters, the other four channels are 12-bit converters
- Two TTL incremental encoder channels
- 20 digital I/O
- 1 three-phase PWM unit, 4 single-phase PWM units

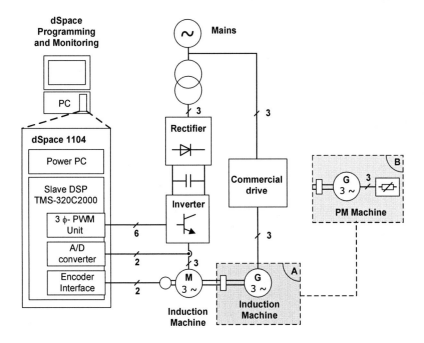

Fig. 5.2   Simplified configuration of the experimental set-ups: A) IM coupled to a synchronous machine acting as a brake machine; B) IM coupled to an IM fed by a commercial drive.

This controller generates the firing pulses for each IGBT, including the suitable dead time to avoid DC-link short-circuit. A standard space phasor modulation with a switching frequency of 5 kHz is considered in this work. Regarding the inverter of the IM of study, it consists of a SKIIPACK intelligent module from Semikron. Further details of the experimental set-up are given in the appendix.

5.2   Faulty IM operating with field oriented control

The experimental results presented next correspond to the variant of the field oriented control that uses the current/angle-model for obtaining the rotor flux angle γ_2 and includes a superimposed velocity control. This control scheme is implemented in the platform based on the Vecon system. For setting the torque level, configuration "B" shown in Fig. 5.2 is used [72].

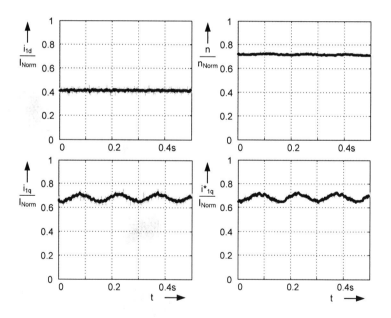

Fig. 5.3   Measured i_{1d}, velocity n, i_{1q} and reference i^*_{1q} of the IM with a single broken rotor bar operating at 1025 min^{-1} near to the nominal load. Set-up B.

For the IM operating without superimposed flux control, at constant velocity and almost nominal torque, the measured flux producing current i_{1d} is not distorted by the fault as shown in Fig. 5.3, which corroborates the theoretical calculations. Conversely, the same figure shows that the torque producing current i_{1q} exhibits a sinusoidal component of twice the slip frequency $2\omega_2$ superimposed to the mean value of i_{1q}, as expected given the inertia of the investigated drive. The reference of the torque producing current i^*_{1q} is distorted by a similar amount by the fault as the measured i_{1q}, since in this case the effects are not masked by the controllers, as explained in 3.2.2. Next, the dependence of the impact of the fault on the torque producing current i_{1q} with the level of torque and with the velocity is investigated in steady-state. Fig. 5.4, Fig. 5.5 and Fig. 5.6 show the measured torque producing current i_{1q} at velocities equal to 731, 1025 and 1410 min^{-1} respectively and with a load torque varying in the range between zero and 90% of the nominal value, which is the maximal torque level allowed with this configuration at its nominal speed. On the one hand, it can be seen that the impact of the fault becomes visible over 30% of the nominal torque and the amplitude of the harmonic component associated to the fault is proportional to the torque level. On the other hand, this behavior is observed at the three tested velocities. Indeed, the impact of the fault on i_{1q} for a similar torque level does not depend on the operating velocity.

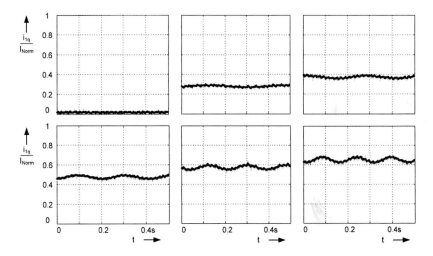

Fig. 5.4   Measured i_{1q} at 731 min^{-1} for the IM with a single broken rotor bar operating under different loads. Set-up B.

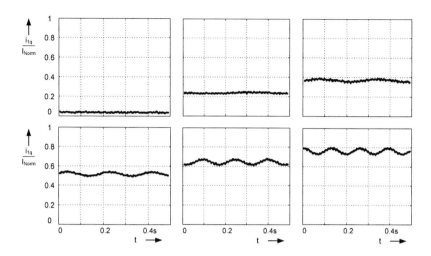

Fig. 5.5   Measured i_{1q} at 1025 min^{-1} for the IM with a single broken rotor bar operating under different loads. Set-up B.

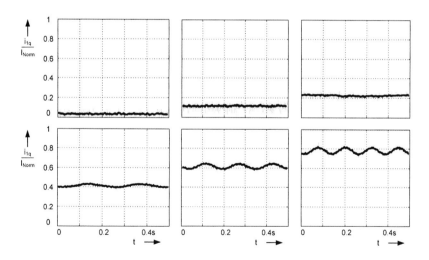

Fig. 5.6   Measured i_{1q} at 1410 min^{-1} for the IM with a single broken rotor bar operating under different loads. Set-up B.

In Fig. 5.7 and Fig. 5.8 it is depicted the resulting measured torque producing current i_{1q} for the healthy IM used as a reference. Operating conditions are similar to those in Fig. 5.5

and Fig. 5.6, respectively. As expected, the measured i_{1q} does not exhibit the characteristic harmonic due to the fault at any torque level.

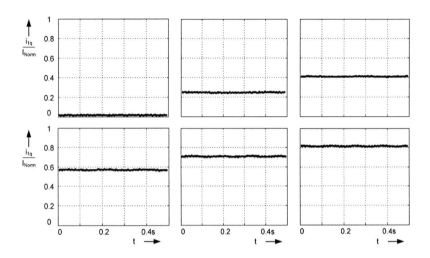

Fig. 5.7   Measured i_{1q} at 1025 min^{-1} for the healthy IM used as reference operating under different loads. Set-up B.

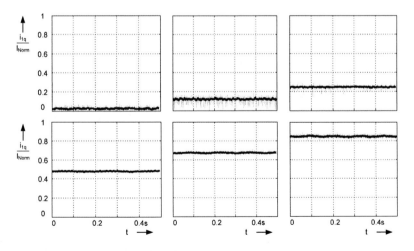

Fig. 5.8   Measured i_{1q} at 1410 min^{-1} for the healthy IM used as reference operating under different loads. Set-up B.

Next, the variant of the field oriented control that uses the stator voltages and currents for obtaining the rotor flux angle φ_2 (see 2.6) and also includes a superimposed velocity control is implemented in the platform based on the dSpace 1104. For setting the torque level, configuration "A" shown in Fig. 5.2 is used. For the following tests, the IM is also loaded near to the nominal point, although some differences must be noted compared to the measurements made with set-up B. The maximal DC-link voltage U_{DC} is in this case 560 V and thus lower than in the set-up B, so the operation near the nominal speed and load is only possible by reducing the proportional controller gain K_{Pn} of the velocity controller and in this way keeping the necessary voltage reserve as low as possible. By inspecting expression (3.13), it can be deduced that the smaller this controller gain is, the smaller the effect on the torque producing current i_{1q}. An additional restriction is introduced by the employed IM which requires maintaining the saturation point and thus enough voltage reserve is only possible at 85% of the nominal velocity under the described conditions.

First, a comparison with the previous scheme that uses the current/angle-model for obtaining the rotor flux angle φ_2 is presented for an equivalent point of operation. This means that the two models are adequately set so that at the examined operating point both control schemes are comparable. Measurements shown in Fig. 5.9 correspond to the faulty IM running at 430 min^{-1}. It can be seen that the impact of the fault on the i_{1q} is visible and similar in both cases.

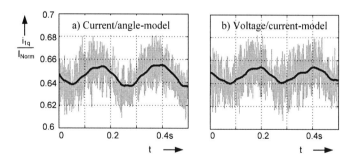

Fig. 5.9   Measured i_{1q} at 430 min^{-1} for the IM with a single broken rotor bar for the two variants of field oriented control with different rotor flux model: a) current/angle model and b) voltage/current model. Original and filtered signals are depicted. Set-up A.

As explained, the impact of the fault is less visible on the torque producing current i_{1q} if compared to the previous velocity controller configuration (see for example Fig. 5.4). Nevertheless, the output of the controllers and the calculated rotor flux $|\underline{\psi}'_2|$ with (2.78) can also be used instead of the measured i_{1q} in order to determine the proposed fault indicator, as shown in Fig. 5.10.

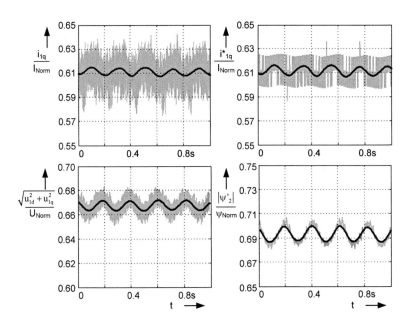

Fig. 5.10   Measured i_{1q}, reference i^*_{1q}, controllers output $\sqrt{u_{1d}^2 + u_{1q}^2}$ and calculated rotor flux $|\underline{\psi}'_2|$ at 1200 min^{-1} for the IM with a single broken rotor bar. Field oriented control performed with voltage/current-model. Original and filtered signals are depicted. Set-up A.

5.3   Faulty IM operating with encoderless field oriented control

The encoderless scheme used in this work is based on the voltage/current-model for the determination of the rotor flux angle, as explained in subchapter 2.6.3. Fig. 5.11 depicts the results of comparative measurements with and without sensor for the rotor position for the similar operating conditions with the faulty machine running at 430 min^{-1} and similar torque level.

It can be seen, that the characteristic harmonic is noticeable and similar in the three evaluated control variants [73].

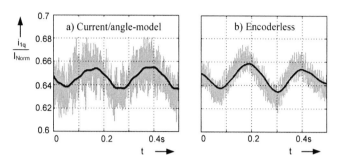

Fig. 5.11   Measured i_{1q} for the IM with a single broken rotor bar running at 430 min^{-1}: a) current/angle-model; b) encoderless. Original and filtered signals are depicted. Set-up A.

5.4   Implementation of the proposed online indicator

The proposed online indicator based on the torque producing current i_{1q} is next presented. Out of the measurements of i_{1q} and twice the slip frequency $2\omega_2$, the amplitude of the component associated to the fault is obtained from applying (4.22) with k=1. It has to be remarked that as explained in the previous subchapter, other variables as the output of the current controllers or the rotor flux amplitude can be advantageous in other cases. On the other hand, the utilization of the stator current is in general preferable since it is always available in commercial drives. Fig. 5.12 shows the coefficient calculated online $|X(n)|$ in this case. As expected, $|X(n)|$ changes with the value of the AC part of the torque producing current, denoted as $i_{1q,\sim}$, indicating the maximum value of the component for each case. The new coefficient $|X(n)|$ corresponding to the new operating point is rapidly obtained. Furthermore, for the case of study $|X(n)|$ can be clearly distinguished from the value of the machine in healthy conditions as shown in Fig. 5.12.c) and d). However, the performance of the indicator during changes in the load level is not adequate. Indeed, its value would mean a more severe fault than a single broken bar in the faulty machine and a false fault alarm for the healthy machine. To partially mitigate this problem, a low pass filter is implemented as indicated in subchapter 4.3. The resulting improved indicator is shown in Fig. 5.13. Finally, the proposed indicator has also been evaluated for the scheme using the voltage/current-model in both cases with and without a sensor for the rotor position for the similar operating conditions shown in Fig. 5.9 and Fig. 5.11 with the faulty machine. A zoom of the AC component of the resulting measured i_{1q} is shown in Fig. 5.14. It can be seen that the proposed indicator is also valid without the necessity of using a sensor for the acquisition of the mechanical position of the shaft.

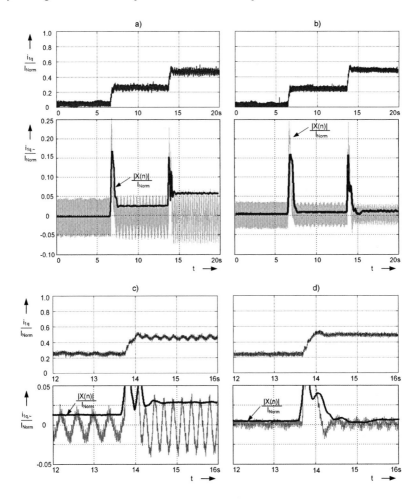

Fig. 5.12   Measured i_{1q}, AC part of the torque producing current $i_{1q,\sim}$ and coefficient calculated online $|X(n)|$ for the IM running at 731 min^{-1}. Transition from 0 to 40 and from 40 to 60% of the nominal load for a) a single broken rotor bar and b) healthy IM used as reference. c) and d) are close views of figures a) and b), respectively. Set-up B.

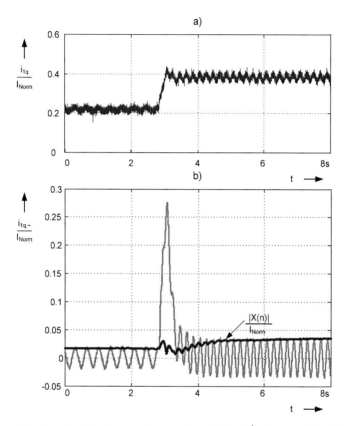

Fig. 5.13   IM with a single broken rotor bar running at 731 min^{-1}: a) measured i$_{1q}$; b) offline calculated coefficient |X(n)| for a transition from 40 to 60% of the nominal load. Set-up B.

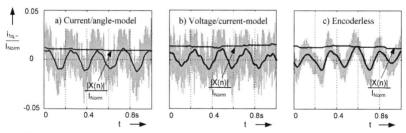

Fig. 5.14   AC part of the torque producing current i$_{1q,\sim}$ and calculated coefficient |X(n)| for the IM with a single broken rotor bar running at 430 min^{-1}: a) current/angle-model; b) current/voltage-mode and c) encoderless. Set-up A.

5.5   Summary of chapter 5

Rotor asymmetries introduce characteristic frequencies in the terminal variables of the motor and they propagate to the rest of control variables. This has been evaluated under different operating conditions and field oriented control schemes with and without sensor for the angular position of the shaft.

Although variables such as the output of the controllers can give a better indication of the fault related components, especially for smaller values of the proportional controller gain K_{Pn} of the velocity controller, the currents were more intensively studied as they are available in any commercial drive. In particular, the impact of the fault becomes visible over 30% of the nominal torque in the torque producing current the amplitude of the harmonic component associated to the fault is proportional to the torque level. Measurements made in a test bench with a healthy and a faulty machine under similar conditions show that a reliable CM based on the torque producing current is possible up to 50% of the nominal load.

The proposed online CM uses an algorithm to determine the amplitude of the characteristic harmonic due to the fault which is based on the slip frequency, a variable which is available in field oriented control. Compared to standard diagnostics methods, this approach permits the detection of the fault related harmonic component even in quasi-stationary conditions, provided that the machine is loaded. The performance of the indicator during a change of load has been improved by means of a low pass filter in order to avoid false alarms.

The fault detection algorithm is evaluated for different variants of a field oriented control including the standard current/angle-model of the IM and the voltage/current-model. This last variant is relevant as it allows the operation of the drive with and without sensor for the angular position of the shaft. The measurements show that the characteristic frequency that indicates a single broken rotor bar position for similar operating conditions is identical in all cases. The proposed algorithm can therefore be used in encoderless control schemes as well.

6 Conclusions

The main objective of this work was the research of diagnostics of rotor asymmetries in the inverter-fed variable speed IM. A new online diagnostics method based on the discrete Fourier transformation has been proposed based on characteristic variables of the control, which works properly even at quasi-stationary conditions with variable frequency.

In the case of IMs, common faults are asymmetries in the squirrel cage as for example broken bars or fractures in the end-rings. This type of fault causes a characteristic harmonic in different variables of the machine that can be used for the purpose of diagnostics. Most of the existing CM techniques have been proposed for machines supplied from the mains. Among them, current monitoring and in particular the application of MCSA is now being extensively used in industry to assess the operational health of rotor cage winding. This technique is based on the frequency domain analysis of the measured stator current, and thus cannot be directly applied to the case of variable speed drive, where the speed varies significantly and thus non-stationary techniques are required. Moreover, in the case of closed-loop inverter-fed IMs, a significant challenge is added to the fault detection due to the action of the controllers, which try to eliminate the disturbances caused by the rotor asymmetry.

A dynamic model to analytically describe squirrel cage IMs including their behavior in both healthy and faulty conditions allows a comprehensive analysis of the effects of the rotor asymmetry. Several variants of this control scheme which differ on the rotor flux model employed and on the utilization or not of angular transducers are analyzed. All schemes constitute state of the art for controlling the IM.

The effects caused by the rotor asymmetries in field oriented controlled IMs are theoretically analyzed in steady-state. In contrast to the cases of line-connected IMs or open-loop inverter drives, for closed-loop inverter-fed drives disturbances in the current are attenuated, provided that the current controllers are properly tuned. In the case of having a superimposed velocity control, the output of the velocity controller constitutes a propagation path to the currents and the torque producing current can give enough information for the diagnostics in the case of medium or low inertia drives. Alternatively, the outputs of the controllers can also be used for the diagnostics. Nevertheless, the torque producing current is preferred in general for the diagnostics because of its availability in commercial drives.

The main algorithms for the digital implementation of Fourier analysis are presented and the most suitable one in terms of less demanding computational efforts and fast update characteristics is selected. These features are important for the implementation. The algorithm uses the angle that is generated directly from twice the slip frequency instead of the time as variable of integration. Then, several signals exhibit the characteristic harmonic caused by the fault and the proposed method produces satisfactory results for all of them. On the one hand, due to its simplicity no extra hardware is required and thus costs are reduced. On the other hand, the fast update characteristics allow the tracking of rapid changes in the input signal. Thus, the proposed algorithm performs properly even at quasi-stationary conditions when the frequency varies. Its performance during transients and dynamic operation has also been discussed. In order to eliminate those false alarms in the detection that may be caused by abrupt changes in the operating conditions of the drive, a simple approach has been proposed which consists on adding a low pass filter to the algorithm of the proposed indicator.

Experimental results are obtained with two different laboratory set-ups. One platform is based on the Vecon system and the second one is based on a prototyping platform dSpace 1104. This second platform facilitates the monitoring of the different variables of the drive and the implementation of the fault detection algorithm. Measurements demonstrate the effectiveness of the proposed encoderless control scheme.

96

7 Abstract

The diagnostics of rotor asymmetries in inverter-fed, variable speed IMs is addressed in this work. In particular, the case of IMs with squirrel cage rotor operating with several variants of the field oriented control has been investigated. The different control schemes differ in the model of the IM being used to obtain the rotor flux angle and also in the use or not of a mechanical sensor to measure the angular position of the rotor.

Rotor asymmetries cause a characteristic harmonic in the different variables of the machine, and proper detection methods are needed in order to avoid further damage to the machine. One of the advantages of diagnostics with field oriented control is that the available variables for control can be also used for CM purposes. Among them, the currents, the output of the controllers or even the flux have been shown to be suitable fault indicators provided certain conditions are met. The characteristic frequency of the fault is in this case known by the control algorithm, as it is proportional to the slip frequency. Last but not least significant, is the fact that no extra hardware is needed, provided that the algorithm for the diagnostics remains simple and compact.

The detection method for rotor asymmetries proposed here is based on the discrete Fourier transformation, but instead of time, it uses the integral of twice the slip frequency. Compared to standard diagnostics methods, this approach permits the detection of the fault related harmonic component even in quasi-stationary conditions, provided that the machine is loaded. It has also been demonstrated that although the indicator may not give the exact value of the harmonic, the sensitivity is enough to detect one single broken bar. For a more incipient fault, for example a partially broken rotor bar, this method may require a threshold for the indicator that can be obtained operating the machine in healthy conditions to differentiate it from possible intrinsic asymmetry due to manufacture.

The fault detection method was evaluated under different operating conditions and field oriented control schemes. Despite variables such as the output of the controller can give a better indication of the fault related components, the currents were more intensively studied because they are available in any commercial drive. It has also been shown that the fault related harmonic is observed in conventional field oriented control with or without a sensor.

8   Zusammenfassung

Rotorunsymmetrien in Asynchronmaschinen verursachen unerwünschte Effekte wie Drehmomentschwankungen und Verluste und sollten frühzeitig erkannt werden, um Folgefehler zu vermeiden. Die vorliegende Arbeit befasst sich deshalb mit der Diagnose von Rotorunsymmetrien in umrichtergespeisten Asynchronmaschinenantrieben variabler Drehgeschwindigkeit. Besonderes Augenmerk der Untersuchung gilt dem Betrieb der Asynchronmaschine mit Käfigläufer in verschiedenen Varianten der feldorientierten Regelung. Im Einzelnen werden der Einfluss des für die Gewinnung des Orientierungswinkels verwendeten Maschinenmodells – mit und ohne Lagesensor- auf Diagnoseproblematik und die notwendigen Algorithmen der Signalverarbeitung betrachtet.

Rotorunsymmetrien verursachen in den Statorströmen einer Asynchronmaschine charakteristische Harmonische, die in vielen früheren Arbeiten zur Fehlererkennung beim Netzbetrieb der Asynchronmaschine herangezogen werden. Im Rahmen der vorliegenden Untersuchung werden geregelte Antriebe variabler Drehgeschwindigkeit betrachtet, in denen nicht nur die Ströme sondern vorteilhaft andere Zeitverläufe der Zustands- und Hilfsvariablen zum Zwecke der Zustandsdiagnose Verwendung finden können. Hierzu gehören unter anderem die Stellgrößen der Stromregler und die Ströme im feldorientierten Koordinatensystem. Ein weiterer Vorteil der geregelten Systeme besteht darin, dass die von der Rotorunsymmetrie verursachte charakteristische Frequenz immer bekannt ist. Sie ist dem Schlupf proportional, der als Zwischenvariable der Regelung stets bekannt ist. Es zeigt sich, dass deshalb die Fehlererkennung im geregelten System einfach, transparent und ohne zusätzliche Sensoren oder Hardware realisierbar ist.

Das hier neu entwickelte Diagnoseverfahren basiert auf der diskreten Fourier-Transformation, die allerdings nicht über eine Zeit- sondern über eine Winkelintegration durchgeführt wird. Der dazu verwendete Winkel wird aus der Integration der charakteristischen Frequenz gewonnen. Im Gegensatz zu den bisher vorgeschlagenen Algorithmen liefert das neue Verfahren dank einer Fensterintegration ein kontinuierliches Signal und kann sogar unter nichtstationären Bedingungen eingesetzt werden.

Auch wenn der neue Algorithmus im Prinzip auf alle Zustandsvariablen der Regelung angewandt werden kann, eignen sich nicht alle für die Fehlererkennung. Neben der Stellgröße der Stromregler und dem Betrag des Rotorflussraumzeigers wurden die Ströme der Maschine im feldorientierten Koordinatensystem als mögliche Fehlerindikatoren intensiv untersucht. Sie liefern genügende Informationen, um die durch einen unterbrochenen Rotorstab hervorgerufene Unsymmetrie bereits bei schwacher Last zu detektieren. Eine Referenzmessung ist dennoch notwendig, um eine Ansprechschwelle zu definieren und fehlerhafte Rotoren von herstellungsbedingten Unsymmetrien unterscheiden zu können.

Die im Labor an zwei verschiedenen Antriebskonfigurationen durchgeführten Messungen beweisen, dass die entwickelte Methode unter unterschiedlichen Betriebsbedingungen und bei verschiedenen Varianten der feldorientierten Regelung mit und ohne Winkelsensor zuverlässig funktioniert.

9 Appendices

9.1 Parameters of the IM (test machine)

Manufacturer:	ABM Greiffenberger Antriebstechnik
Nominal power:	2.2 kW
Nominal voltage:	380 V
Nominal current:	5.1 A, star-connection
Nominal frequency:	50 Hz
Nominal speed:	$1410 \, \text{min}^{-1}$
Nominal power factor:	0.8
Stator resistance (20°C) R_1:	3.37 Ω
Rotor resistance (20°C) R'_2:	2.34 Ω
Stator leakage inductance $L_{1\sigma}$:	11.4 mH
Rotor leakage inductance $L'_{2\sigma}$:	12.9 mH
Mutual inductance L_h:	286.2 mH

9.2   Set-up A

In this configuration, the load torque level was set by using an IM as shown in Fig. I driven by a commercial drive with a chopper module. Control and signal processing are implemented in a DSP add-on card (DSPACE DS1104), as shown in Fig. II. The voltage source inverter and the corresponding drives are integrated in an intelligent power module of SEMIKRON with the following characteristics:

- ○ SK integrated Intelligent Power 6-pack. SKiiP 232GD120-313CTVU
- ○ V_{CES} = 1200 V
- ○ Ic   = 200 A
- ○ t_{dead} = 3 μs
- ○ f_{MAX} = 20 kHz

The DC-link voltage U_{DC} can be modified, being the maximal U_{DC} employed in this work equal to 560 V. The parameters of the IM used as a load machine are:

Manufacturer:	System Antriebstechnik, Dresden GmbH
Nominal power:	5.5 kW
Nominal voltage:	360 V, line-to-line
Nominal current:	12.7 A, star-connection
Nominal frequency:	52 Hz
Nominal speed:	1500 min^{-1}
Nominal torque:	35 Nm
Nominal power factor:	0.81
Pole pairs p:	2

Fig I. Test bench of the set-up A.

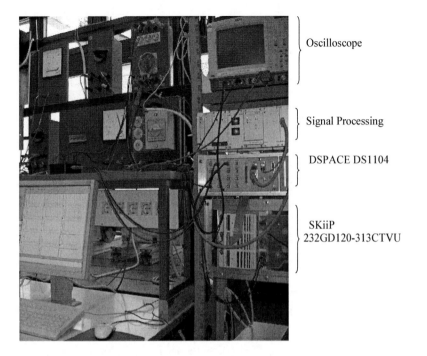

Oscilloscope

Signal Processing

DSPACE DS1104

SKiiP
232GD120-313CTVU

Fig II. Frontal view of the rack corresponding to set-up A.

9.3   Set-up B

In this configuration, a permanent magnet synchronous machine with a variable resistor was used to set the load level.

The parameters of three-phase permanent magnet servo motor used as a load machine are:

Manufacturer, motor type: ABB Asea Brown Boveri, SDM101-008N3-115

Nominal voltage: 360 V, star-connection

Standstill motor current: 5.3 A, star-connection

Maximal impulse current: 27 A

Nominal speed: $3000\ \mathrm{min^{-1}}$

Standstill torque: 8.3 Nm

Inertia: $0.89 \cdot 10^{-3}\ \mathrm{kgm^2}$

In this case, control and signal processing were implemented in a platform using a Vecon based control system which is especially designed for electric drive applications of high dynamics [60]. The DC-link voltage U_{DC} is controlled to a fix value of 670 V. The inverter uses a space vector PWM, the switching frequency is 5 kHz and the sampling time is 100 µs. Further details of the control and signal processing board are given in [60]. In this case, the sinusoidal analogue signals of a 2048 pulse incremental encoder are used. By analogue interpolation, this resolution can be improved to approx. 0.5 Mio increments per revolution [46].

9.4 Normalization

The following table lists the normalization values for the variables that have been used in this work. They can be obtained from the nominal per phase stator voltage U_{phN} and nominal per phase stator current I_{phN} and the nominal angular frequency ω_{sN}. For the normalization of voltage, current and flux linkages, their corresponding maximal values are considered.

Normalization values:

Voltage
$$U_{Norm} = \sqrt{2}U_{phN}$$

Current
$$I_{Norm} = \sqrt{2}I_{phN}$$

Frequency
$$\omega_{sN}$$

Flux linkage
$$\psi_{Norm} = \frac{\sqrt{2}U_{phN}}{\omega_{sN}}$$

10   References

[1]   IAS Motor Reliability Working Group.
        *Report of Large Motor Reliability Survey of Industrial and Commercial Installa-
        tions, Part I.*
        IEEE Trans. on Industry Applications, vol. IA-21 (1985) no. 4, pp. 853-872

[2]   Thorsen, O.V.; Dalva, M.
        *A survey of faults on induction motors in offshore oil industry, petrochemical in-
        dustry, gas terminals and oil refineries.*
        IEEE Trans. on Industry Applications, vol. 31 (1985), no. 5, pp. 1186-96

[3]   Bonnett, A.H.; Soukup, G.C.
        *Cause and analysis of stator and rotor failures in three-phase squirrel-cage induc-
        tion motors.*
        IEEE Trans. on Industry Applications, vol. 28 (1992), no. 4, pp. 921-937

[4]   Durocher, D.B.; Feldmeier, G.R.
        Predictive versus preventive maintenance.
        IEEE Industry Applications Magazine, vol. 10 (2004), no. 5, pp. 12-21

[5]   Briz, F.; Degner, M.W.; Díez, A.B.; Guerrero, J.M.
        *Online diagnostics in inverter-fed induction machines using high-frequency signal
        injection.*
        IEEE Trans. on Industry Applications, vol. 40 (2004), no. 4, pp. 1153-61

[6]   Filippeti, F.; Bellini, A.; Franceschini,G.; Tassoni,C.
        Closed-loop control impact on the diagnosis of induction motor faults.
        IEEE Trans. on Industry Applications, vol. 36 (2000), no. 5, pp. 1318-29

[7]   Frank, P.M.
        Diagnoseverfahren in der Automatisierungstechnik.
        Automatisierungstechnik, vol. 42 (1994), no.2, pp. 47-64

[8]   Frank, P.M.; Ding, S.X.; Marcu,T.
        Model-based fault diagnosis in technical processes.
        Trans. of the Inst. of Measurement and Control, vol. 22 (2000), no. 1, pp. 57-101

[9]   Isserman, R.
        Fault-diagnosis systems.
        Springer Berlin, 2005

[10]   Schuisky, W.
        *Brüche im Kurzschlußkäfig eines Induktionsmotors und ihre Einflüsse auf das
        Verhalten des Motor.*
        Archiv für Elektrotechnik, vol. 35 (1941), no. 5, pp. 287-298

[11]   Hiller, U.M.
        *Einfluss fehlender Läuferstäbe auf die elektrischen Eigenschaften von Kurz-
        schlussläufer-Motoren.*
        ETZ-Archiv, vol. 83 (1962), pp. 94-97

[12]   Jordan, H.; Weis, M.
         Asynchronmaschinen.
         Vieweg,1969

[13]   Rogge, D.; Seinsch, H.O.
         Erkennung und Überwachung von elektrischen Läuferunsymmetrien in Käfigläufern.
         ETZ-Archiv, vol. 3 (1981), pp. 39-45

[14]   Williamson, S.; Smith, A.C.
         Steady-state analysis of 3-phase cage motors with rotor-bar and end-ring faults.
         IEE Proceedings, vol. 129 (1982), Pt. B, no. 3, pp. 93-100

[15]   Deleroi, W.
         Der Stabbruch im Käfigläufer eines Asynchronmotors. Teil 1: Beschreibung mittels Überlagerung eines Störfeldes.
         Archiv für Elektrotechnik, vol. 67 (1984), pp. 91-99

[16]   Kliman, G.B.; Koegl, R.A.; Stein, J.; Endicott, R.D.; Madden, M.W.
         Noninvasive detection of broken bars in operating induction motors.
         IEEE Trans. on Energy Conversion, vol. 3 (1988), no. 4, pp. 873-879

[17]   Elkasabgy, N.M.; Eastham, A.R.; Dawson, G.E.
         Detection of broken bars in the cage rotor on an induction machine.
         IEEE Trans. on Industry Applications, vol. 28 (1992), no.1, pp. 165-171

[18]   Deng, X.
         Detection of rotor faults on induction motors by investigating the flux linkage of the stator winding.
         Dissertation, University of Helsinki, 1994

[19]   Toliyat, H.A.; Lipo, T.A.
         Transient analysis of cage induction machines under stator, rotor bar and end ring faults.
         IEEE Trans. on Energy Conversion, vol. 2 (1995), no.1, pp 241-247

[20]   Penman, J; Stavrou, A.
         Broken rotor bars: their effect on the transient performance of induction machines.
         IEE Proceeding Electrical Power Applications, vol. 143 (1996), no. 6, pp. 449-457

[21]   Bellomi, A.
         Beitrag zur Analyse statorunsymmetrischer Drehfeldmaschinen.
         Dissertation, Tech. Hochsch. Karlsruhe, Germany, 1987

[22]   Munoz-García, A.; Lipo, T.A.
         Complex vector model of the squirrel cage induction machine including instantaneous rotor bar currents.
         IEEE Trans. on Industry Applications, vol. 35 (1999), no. 6, pp. 1132-1340

[23]   Tavner, P.J.; Penman, J.
         Condition Monitoring of Electrical Machines.
         Research Studies Press Ltd., 1st Edition, England, 1987

[24]   Früchtenicht, F.; Pittius, E.; Seinsch, H.O.
        A diagnostic system for three-phase asynchronous machines.
        Electric Machines and Drives Conference 1989, Proc., pp. 163-171

[25]   Cabanas, M.F.; Melero, M.G.; Orcajo, G.A; Rodríguez, J.M.C; Sariego, J.S.
        Técnicas de mantenimiento y diagnóstico en máquinas eléctricas rotativas.
        Marcombo Boixareu Editores, Spain, 1998

[26]   Cabanas, M.F.; Glez, F.P.; Gonzalez, M.R.; Melero, M.G.; Orcajo, G.A.; Rojas,
        C.H.
        *A new on-line method for the early detection of broken rotor bars in asynchronous
        motor working under arbitrary load conditions.*
        IEEE Industry Applications Society Annual Meeting 2005, Proc., vol. 1, pp. 662-9

[27]   Cameron, J.R.; Thomson, W. T.; Dow, A.B.
        *Vibration and current monitoring for detecting airgap eccentricity in large induc-
        tion motors.*
        Electric Power Applications, IEE Proceedings B, vol. 133 (1986), no.3, pp. 155-63

[28]   Pöyhönen, S.
        *Support vector machine based classification in condition monitoring of induction
        motors.*
        Dissertation, Helsinki University of Technology. Espoo, 2004

[29]   Thomson, W.T.
        *On-line current monitoring- The influence of mechanical load or a unique rotor
        design on the diagnosis of broken bars in induction motors.*
        International Conference on Electrical Machines 1992, Proc., pp. 1236-1240

[30]   Schoen, R.R.; Lin, B.K.; Habetler, T.G.; Schlag, J.H.; Farag, S.
        *An unsupervised, on-line system for induction motor fault detection using stator
        current monitoring.*
        IEEE Trans. on Industry Applications, vol. 31 (1995), no. 6, pp. 1280-6

[31]   Filippeti, F.; Bellini, A.; Franceschini, G.; Tassoni, C.; Kliman, G.B.
        *Quantitative evaluation of induction motor broken bars by means of electrical
        signature analysis.*
        IEEE Trans. on Industry Applications, vol. 37 (2001), no. 5, pp. 1248-55

[32]   Henao, H.; Razik, H.; Capolino, G.A.
        *Analytical approach of the stator current frequency harmonics computation for
        detection of induction machine rotor faults.*
        IEEE Transactions on Industry Applications, vol. 41 (2005), no. 3, pp. 801-907

[33]   Cardoso, A.J.M.; Cruz, S.M.A.; Carvalho, J.F.S.; Saraiva, E.S.
        *Rotor cage fault diagnosis in three-phase induction motors, by Park's vector ap-
        proach*
        EEE Industry Applications Society Annual Meeting 1995, Proc., vol. 1, pp.642-6

[34]   Schoen, R.; Habetler, T. G.
        Effects of time-varying loads on rotor fault detection in induction machines.
        IEEE Trans. on Industry Applications, vol. 4 (1995), no.31, pp. 900-6

[35]   Wu, L.; Habetler, T.G.; Harley, R.L.
         A review of separating mechanical load effects from rotor faults detection in induction motors.
         Diagnostics for Electric Machines, Power Electronics and Drives, 2007, IEEE International Simposium on, 2007, pp. 221-225

[36]   Watson, J.F; Paterson, N.C.
         Improved techniques for rotor fault detection in three-phase induction motors
         IEEE Industry Applications Society Annual Meeting 1998, Proc., vol. 1, pp.271-277

[37]   Benbouzid, M.E.H.; Kliman, G.B.
         What stator current processing-based technique to use for induction motor rotor faults diagnosis?
         IEEE Trans. On Energy Conversion, vol. 18 (2003), no. 2, pp. 238-244

[38]   Kral, C.; Pirker, F.;
         Vienna Monitoring Method- Detection of faulty rotor bars by means of a portable measurement system
         International Conference on Electrical Machines 2000, Proc., vol. 2, pp. 873-77

[39]   Filippetti, F.; Franceschini, G.; Tassoni, C; Vas,P.
         AI techniques in induction machines diagnosis including the speed ripple effect.
         IEEE Trans. on Industry Applications, vol. 34 (1998), no.1, pp. 98-108

[40]   Wieser, R. S.; Schagginger, K. C.; Kral, Ch.; Pirker, F.
         The integration of machine fault detection into an indirect field oriented control induction machine drive control scheme. The Vienna Monitoring Method.
         IEEE Industry Applications Society Annual Meeting 1998, Proc., vol. 1, pp. 278-285

[41]   Wieser, R. S.; Schagginger, K. C.; Kral, Ch.; Pirker, F.
         The Vienna induction machine monitoring method; on the impact of the field oriented control structure on real operational behavior of a faulty machine.
         Annual Conference of the IEEE Industrial Electronics Society 1998, Proc., pp. 1544-1549

[42]   Mirafzal, B.; Demerdash, N.A.O.
         Effects of load magnitude on diagnosing broken bar faults in induction motors using the pendolous oscillation of the rotor magnetic field orientation.
         IEEE Trans. on Industry Applications, vol. 41 (2005), no. 3, pp. 771-83

[43]   Späth, H.
         Elektrischen Maschinen.
         Springer Verlag Berlin, Heidelberg, New York. 1973

[44]   Jordan, H.; Klima, V.; Kovacs, K.P.
         Asynchronmaschinen.
         Vieweg, Braunschweig, 1975

[45]   Richter, R.
         Elektrischen Maschinen. Vierter Band. Die Induktionsmaschine.
         Verlag Birkhäuser, Basel, Stuttgart, 1954

[46]   Leonhard, W.
         Control of electrical drives.
         Springer Verlag Berlin, Heidelberg, New York, 2001

[47]   Späth, H.
         Steuerverfahren für Drehstrommaschinen. Theoretische Grundlagen.
         Springer Verlag Berlin, Heidelberg,1983

[48]   Hasse, K.
         Zur Dynamik drehzahlgeregelter Antriebe mit Stromrichtergespeisten Asynchron-kurzschlussläufermaschinen.
         Dissertation, Tech. Hochsch. Darmstadt, Germany, 1969

[49]   Blaschke, R.
         Das Verfahren der Feldorientierung zur Regelung der Drehfeldmaschine.
         Dissertation, Tech. Univ. Braunschweig, Germany, 1974

[50]   Quang, N.P.; Dittrich, J.A.
         Praxis der feldorientierten Drehstromantriebsregelungen.
         Expert Verlag, 1999

[51]   Holtz, J.
         Perspective of sensorless AC Drive Technology.
         International Conference & Exhibition for Power Electronics, Intelligent Motion and Power Quality 2005, Proc., 80-87

[52]   Bauer, F.; Heining, H.D.
         Quick response space vector control for a high power three-level-inverter drive system.
         Electrical Engineering (Archiv für Elektrotechnik), vol. 74 (1990), no. 1, pp. 53-9

[53]   Holtz, J.
         Sensorless control of induction machines- with or without signal injection?
         IEEE Trans. On Industrial Electronics, vol. 53 (2006), no. 1, pp. 7-30

[54]   Schauder, C.
         Adaptive speed identification for vector control of induction motors without rotational transducers.
         IEEE Trans. On Industry Applications, vol. 28 (1992), no. 5, pp. 1054-61

[55]   Kubato, H.; Matsuse, K.; Nakano, T.
         DSP based speed adaptive flux observer of induction motor.
         IEEE Trans. On Industry Applications, vol. 29 (1993), no. 2, pp. 344-48

[56]   Lascu, C.; Boldea, I.; Blaabjerg, F.
         Very low speed sensorless variable structure control of induction machine drives without signal injection.
         IEEE Int. Electric Machines and Drives Conf. 2003, Proc., vol. 3, pp. 1395-1401

[57]   Kang, S.J.; Kim, J.M.; Sul, S.K.
         Position sensorless control of synchronous reluctance motor using high frequency current injection
         IEEE Trans. On Energy Conversion, vol. 14 (1999), pp. 1271-1275

[58]   Briz, F.; Degner, M.W.; Díez, A.B.
         Dynamic operation of carrier-signal-injection-based sensorless direct field-oriented AC drives.
         IEEE Trans. on Industry Applications, vol. 36 (2000), no. 5, pp. 1360-68

[59]   Degner, M.W.; Lorenz, R.D.
         Using multiple saliencies for the estimation of flux, position and velocity in AC machines.
         IEEE Trans. On Industry Applications, vol. 34 (1998), vol. 34, no. 5, pp. 1097-04

[60]   Kilthau, A.
         Drehgeberlose Regelung der Synchronen Reluktanzmaschine.
         Dissertation Universität Siegen, 2002

[61]   Föllinger, O.
         Regelungstechnik-Einführung in die Methoden und ihre Anwendung.
         8. Auflage, Hüthig Buch Verlag Heidelberg, 1994

[62]   Serna Calvo, E.T.; Pacas, J.M.
         Limitations on detecting rotor asymmetries from the measured currents in closed loop operation.
         IEEE International Symposium on Industrial Electronics 2007, Proc., pp. 1311-16

[63]   Pfaff, G.; Meier, Ch.
         Regelung elektrischer Antriebe II.
         Verlag München Wien, 1992

[64]   Lutz, H.; Wendt, W.
         Taschenbuch der Regelungstechnik.
         Harri Deutsch Verlag, 1998

[65]   Klingen, B.
         Fouriertransformation für Ingenieur- und Naturwissenschaften.
         Springer Verlag Berlin, 2001

[66]   Meyer, M.
         Signalverarbeitung. Analoge und digitale Signale, Systeme und Filter.
         Friedr. Vieweg., 1998

[67]   Goertzel, G.
         An algorithm for the evaluation of finite trigonometric series
         Am. Math. Month. 65 (1958), pp. 34-35

[68]   Halberstein, J.H.
         Recursive, complex Fourier analysis for real-time applications.
         Proc. IEEE, vol. 54 (1966), pp. 903

[69]   Oppenheim, A.V.; Schafer, R.W.
         Digital signal processing.
         Prentice-Hall International, Inc. London, 1975

[70]   Rabiner, L., Gold, B.
         Theory and application of digital signal processing.
         Upper Saddle River, NJ: Prentice Hall, 1975

[71]   Aravena, J.L.
         Recursive moving window DFT algorithm
         IEEE Trans. Comput., vol. 39 (1990), pp. 145-148

[72]   Serna Calvo, E.T.; Pacas, J.M.
         Detection of rotor faults in field oriented controlled induction machines.
         IEEE Industry Applications Conference 2006, Proc., vol. 5, pp. 2326-2332

[73]   Serna Calvo, E.T.; Pacas, J.M.
         *Diagnosis of rotor asymmetries in induction machines with different field oriented
         schemes.*
         Annual Conference of the IEEE Industrial Electronics Society 2008, Proc., pp.
         1143-1148